Builder's Guide to Wells and Septic Systems

Builder's Guide to Wells and Septic Systems

R. Dodge Woodson

McGraw-Hill

New York San Francisco Washington, D.C. Auckland Bogotá
Caracas Lisbon London Madrid Mexico City Milan
Montreal New Delhi San Juan Singapore
Sydney Tokyo Toronto

McGraw-Hill

*A Division of The **McGraw·Hill** Companies*

©1996 by **R. Dodge Woodson**.
Published by The McGraw-Hill Companies, Inc.

Printed in the United States of America. All rights reserved. The publisher takes no responsibility for the use of any materials or methods described in this book, nor for the products thereof.

hc   1 2 3 4 5 6 7 8 9 0 DOC/DOC 9 0 0 9 8 7 6
pb   3 4 5 6 7 8 9 0 DOC/DOC 9 0 2 1 0 9 8

Library of Congress Cataloging-in-Publication Data
Woodson, R. Dodge (Rodger Dodge), 1955—
    Builder's guide to wells and septic systems / by R. Dodge Woodson
      p.   cm.
    Includes index.
    ISBN 0-07-071782-6 (hardcover)   ISBN 0-07-071839-3 (paperback)
    1. Wells. 2. Septic tanks. I. Title
  TD405.W66   1996
  628.1'14—dc20   96-4796
                                             CIP

McGraw-Hill books are available at special quantity discounts to use as premiums and sales promotions, or for use in corporate training programs. For more information, please write to the Director of Special Sales, McGraw-Hill, 11 West 19th Street, New York, NY 10011. Or contact your local bookstore.

Acquisitions editor: April Nolan
Editorial team: Tom Begnal, Editor
              Susan W. Kagey, Managing Editor
              Lori Flaherty, Executive Editor
Production team: Katherine G. Brown, Director
              Wanda S. Ditch, Desktop Operator
              Linda L. King, Proofreading
              Joann Woy, Indexer
Design team: Jaclyn J. Boone, Designer
              Katherine Lukaszewicz, Associate Designer   0717826
                                                           GEN1

To my Mom.
As I write this, she's in the hospital, but she's a fighter.
By the time she reads this, she will have beaten the battle.
Mom, I love you.

Acknowledgments

I'd like to acknowledge Kimberley, my wife, for her never-ending support. Afton and Adam, my children, provide me with the inspiration to write. My father, Woody, has always been there when I needed him. I thank all of these very special people for making my life more enjoyable.

Three companies assisted in this project by providing art work. I'd like to thank Amtrol, Inc., A.Y. McDonald Mfg. Co., and Goulds Pumps, Inc. for their generous contributions of drawings that make the text of this book easier to understand.

Contents

Introduction

So you want to know more about wells and septic systems. Well, you've come to the right place. Your decision to pick up this book could make your building business much more profitable. Whether you are someone who has never built a house that needed a well or septic system or a seasoned contractor who regularly deals with rural property, you are going to love this book.

I'm in a unique position to help answer your questions about wells and septic systems. First, I'm a master plumber who has worked with hundreds of wells, pumps, and septic systems. In addition to my plumbing credentials, I'm also a builder. I've built as many as 60 single-family homes a year, many of them served by wells and septic systems. My construction experience spans more than two decades.

Would you like to make more money from the houses that you build? Would you want to avoid losing thousands of dollars on jobs? Of course you would, and this book can help. I've made a lot of mistakes over the years, and you can learn from them. My investment in costly on-the-job lessons is available to you here, in an affordable and easy-to-use book.

I'm glad you decided to pick up this valuable tool. Maybe you don't think of books as tools, but this one is going to be a very valuable addition to your tool box or desk drawer. You might have no desire to take a hands-on position when it comes to wells or septic systems. That's fine. You don't have to. The information and instructions in this book allow you to take either a white-collar approach or a very active, hands-on approach. It's up to you. The advice is here for either path.

What can you gain from this book? More than you can imagine. Let me give you a brief overview. Chapter 1 gives you a broad-brush view of how money is lost when dealing with wells and pump systems. Chapter 2 provides this same type of information for septic systems.

Chapter 3 shows you what to look for when you are doing site inspections for water wells. You can determine whether a drilled well is likely to be needed, or if a less expensive, shallow well can work.

When you get to Chapter 4, you are going to find plenty of helpful information on septic systems. Don't build a house with a septic system before reading this chapter. Critical to success as a rural builder is knowing what to look for when walking the land where a septic system is needed.

How many times have you lost money after bidding jobs poorly? It happens to all of us from time to time. But you can reduce the odds of losing money by reading Chapter 5.

Septic systems involve some technical preparation. You must be aware of code requirements. It's also necessary to have soil studies done and septic designs drawn. Chapter 6 walks you through this phase of the job.

Drilled wells are one of the most dependable private water sources available. These wells aren't cheap, but they are often needed. Chapter 7 provides some important advice on drilled wells.

Shallow wells are much less expensive than drilled wells. These wells are not always suitable, but they are a good value when they can be used. Geological conditions affect the use of bored and dug wells. Read Chapter 8 to discover the mysteries of shallow wells.

Would you want to use your water well to keep your lawn or garden irrigated? Irrigation is needed most in dry seasons, which also happens to be the time of year when wells are at their lowest level. Drawing too much water from a well during dry times can leave a house without water. Alternative water sources can be the answer to this problem. Whether irrigating, watering livestock, or using water for domestic purposes, alternative water sources deserve your attention. See Chapter 9 for details.

Chapter 10 gives you the goods on the most common and least expensive types of septic systems. If a more complex septic system is needed, you can find answers in Chapter 11. These two chapters provide a good understanding of all typical septic systems.

Gravity-type septic systems cost much less to install than pump systems. Chapters 12 and 13 describe all that you need to know about these two types of septic systems.

Saving money, and making more money, are subjects most contractors find very interesting. Who wouldn't like to make or save more money? Chapters 14 and 15 show you how it's done. Flip to these chapters and see for yourself.

Problems come up from time to time. This is true of most things in life. As a builder who installs wells, you are likely to get some late-night phone calls from complaining customers. Chapter 16 can help you get through these times. The troubleshooting tips in this chapter are great. And, Chapter 17 does the same for septic systems.

I've never met a contractor who likes cost overruns. Every builder wants to stay on budget and see handsome profits. Unfortunately, cost overruns are common in the building industry. Is there a way to avoid them? You bet. You can keep yourself on budget in a number of ways, and Chapter 18 explains them to you.

1

Water wells can be money pits

Water wells can be money pits, but they don't have to be. When constructing houses in rural areas, builders are often required to install wells as private water sources. This, in itself, is no big deal. But, if you don't know what to look for, you can lose a lot of money installing wells and their pump systems.

If you have built houses that required wells, you have some experience with what goes on when bidding jobs with wells. A lot of builders never have to deal with well installers. First-time builders and contractors who are not accustomed to working with wells are at risk when it comes to wells.

How much do you already know about wells? Can you tell me what a driven well is? Do you know the difference between a drilled well and a bored well? Should you agree to a per-foot price when having a well installed? These are just some of the questions that are going to be answered in this book.

Three main types of wells are used in modern building practices: drilled wells, bored wells, and driven wells. Drilled wells and bored wells are the two most common types of residential wells.

Are there alternatives to a well when a private source of water is needed for a house? Sure there are. Natural springs are one type of alternative water source. There are many others.

Water that is safe to drink is called *potable water*. However, not all alternative sources produce water that is potable. Does this rule alternative water sources out of consideration altogether? No, because not all water must be safe for domestic use. Watering livestock, irrigating lawns and gardens, running water-based heat pumps, and other needs can be handled by non-potable water.

How much
do you need to know?

Builders don't need to know all the technical aspects related to the installation of a well or pump system. However, to bid jobs with confidence and to supervise the work of your subcontractors, you should
have a better than average understanding of how wells are installed
and used.

A well is much more than a hole in the ground that provides water. What you don't know can hurt you. It can hurt you financially and
damage your reputation. An ignorant builder is an easy mark for an
unscrupulous well installer. You can lose a lot of money if you don't
know what to expect when having a well installed.

The fact that you are reading this book is a good sign. It indicates
that you want to be a responsible, profitable builder. Not all builders
share your willingness to learn. This is to your advantage. Once you
know enough about wells and their pumping systems, you can talk
with authority to potential customers. This can be your competitive
edge in a tight bidding contest. Let's consider this fact a little more
closely.

Assume that you are a prospective home buyer. You want to have
a custom home built on your rural land. A well is needed to supply
the house with water. You've never built a house before and don't
have any friends in the area who have built recently, so you're not
sure how to find the best builder. Being an intelligent person, you go
through the normal channels to screen potential builders. After
checking many sources, you've narrowed the field down to four
builders.

After selecting the four builders to bid your job, you go out and
look over some houses they have built. One of the four builders
doesn't quite feel right to you. It's nothing that you can really put
your finger on, but the chemistry just isn't right. You weed that
builder out of contention. Now it's down to three.

You can't decide which of the three remaining builders you are
the most comfortable with. All of them seem about equal, so you assume price is going be the main factor in your choice.

When the builders present their proposals, one of them is a dis-
appointment. The presentation lacks professionalism and the builder
seems too busy to offer a detailed bid or to answer your questions. A
quick decision is made to eliminate this builder, who also happens to
be the high bidder. Now you are left with two builders. Ironically,

their prices are so close that money is hardly a worthy yardstick for measuring the two contractors. You are perplexed.

Which builder to select?

Both builders present their proposals in person. You like this fact. Unlike the unprofessional builder, these contractors followed your bidding instructions to the letter. The job could go to either one. Then you realize that there are still a few remaining questions and concerns. You call both builders and schedule appointments to talk with them further.

Having never lived in the country before, you are concerned about using a well for your drinking water. Is well water really safe to drink? What types of problems might arise from living with well water? For that matter, what type of wells do people in the country use? Many questions about wells race through your mind.

You meet with the first builder and present your questions. All the basic building questions are answered flawlessly. However, when you start asking questions about your new well, the builder stumbles and can't answer many of them. You are able to find out that your house probably requires a drilled well and that it could be 200 feet deep or deeper. But, most of the more pointed questions are beyond the builder's scope of knowledge. The best the builder can do is offer to let you talk to the well installer and plumber who are going to put your well system together. This isn't a bad option, and you respect the builder for being honest.

Before leaving, the builder recommends the names of some local authorities who might be able to help you with your well questions. The builder makes arrangements for you to meet with the well-system subcontractors. After the builder leaves, you make a few phone calls and talk with local authorities to establish some background information on wells. Fortunately, you get a number of your questions answered.

The next day, the second builder shows up to go over any questions and concerns you might have. Since you had prepared a list of questions for the first builder, you use this same list with the second builder. Just like the first contractor, this builder glides through the basic building questions with ease. The answers coincide with those from the first builder, so you feel confident that they are valid.

When you get to the bottom of your list and start asking questions about the well, you expect the builder to either bluff you with nondescriptive answers or to refer you to experts in the field of wells.

It amazes you when the builder is able to answer your questions quickly and without referring to reference material.

The builder explains the various types of well options and confirms that you probably need a drilled well. But, the builder goes much further, even beyond the scope of your questions. You're told the advantages of using a submersible pump instead of a two-pipe jet pump. The purpose and importance of pressure tanks is discussed. Alternative water sources are explained to you as an option for your horses.

This builder goes on and on about water quantity and quality. Installation methods are explained to you. The builder tells you how the well is disinfected and tested before you are allowed to use it. This puts your mind at ease. Many other aspects of well systems are explained, including the possible need for water treatment equipment. You are quickly seeing a builder who has a lot of experience with wells, and this makes you feel secure.

When the builder is done, you compare the answers with those that you got by telephone from local authorities. They jive. This builder obviously knows about well systems. No longer are you confused about which builder should be awarded your job. Both builders made good impressions, but the second builder has instilled confidence in you that the other builder couldn't. The simple fact that this particular builder had a depth of knowledge about wells is enough to help you make the choice.

Which of these builders would you rather be? Do you want to be the one who gets the job? If you do, keep reading this book. By the time you finish it, you are going to be in a position to separate yourself from much of the competition. The ability to answer customer questions about wells really can make a big difference in a bidding war.

What do you need to know?

What do you need to know about wells? There is not a lot that you absolutely must know in order to build houses that use wells. You can rely on subcontractors to get you through the process. But, if you want to be in control and see higher profits, there is much that you should know. Let me give you some examples.

When you walk a piece of land with a prospective home buyer, do you know what characteristics to look for in terms of well installation? Do you have any idea whether or not a shallow well can be used? You can't determine everything you need to know about well selection by simply looking at a piece of land, but the land can give you lots of signals.

You need to know what to look for when you do a site inspection. For example, the presence of bedrock indicates that a drilled well is going to be needed. This is important information, since drilled wells cost much more than bored wells. Knowing the legal distance between a well and a septic system can tip you to problems with certain pieces of land. Walking a building lot might reveal that its size and layout is going to make a well and septic installation difficult. In order to make the lot buildable, some changes might be needed, either in the proposed house design or the house placement.

Buildable lots

Buildable lots are needed to build houses. A lot might consist of a one-acre parcel or acres of rolling land. You know that the land chosen must be buildable, but what is the definition of a buildable lot? Basically, it is one where a house can be built in compliance with local codes and zoning requirements. However, it doesn't necessarily mean that the specific house you or your customer wants to build can be constructed on the lot. This is important, so think about it.

Buying a lot that is deemed buildable doesn't guarantee that you can build what you want to build. Neither does it assure you that your construction can take place in the location of your choice. Most houses that use well water also use septic systems. When wells and septic systems are both needed, certain regulations come into play. Every jurisdiction I know has rules pertaining to the proximity of wells to septic systems. These rules can ruin your plans. Let's look at an example.

Assume that you are a spec builder. You've stumbled onto a great building lot. The location requires a septic system and a well. Soils tests have already been done, and the land is approved for a conventional septic system. You are excited about the find, so you move quickly to make the purchase.

Even though the land is along a rural road, it's in a neighborhood of large homes. Surrounding properties display houses with three to four bedrooms and an average of 2200 square feet of living space. On tax maps, the lot size proves to be similar to other lots in the area. Everything looks good.

You plan to build a two-story Colonial home on the property. There is going to be a two-car, attached garage and an attached mud room connecting the garage to the kitchen. This won't be a huge house, but it is going to check in at more than 2000 square feet. You've run speculative numbers, and the deal looks like a real money-maker. Then it happens.

In your haste to secure the building lot, you neglected to do all of the preliminary site planning. Your broker covered the bases by making certain that the lot was buildable, but you didn't take the time to lay the house out on the land, along with the proposed septic system, to see how everything fit. Now you've run into a problem as you prepare to apply for a construction loan.

When you laid out the septic system and well location, you noticed that there was not enough room to place the house of your choice on the lot in a position that is going to be appealing. In fact, you might not even be able to get the house on the lot without omitting the garage. All of the surrounding properties have garages, so building a spec house without one would be very risky. You've got a mess on your hands.

You now own a building lot, but you can't build the house you wanted or locate it as planned. Since you were planning a two-story home, there's not much you can do to make the footprint of the foundation smaller without losing square footage. Maybe you can switch to a Cape Cod design with an ell wing, but that's something you would prefer not to do. In any event, you have a building lot and don't know what to do with it. Between zoning set-backs and the distance required between the well and septic system, you simply don't have a lot of building room.

What created this problem? Other lots of the same size are not affected in this way. You have a problem because the back edge of the building lot does not have soil consistency to support a standard septic system. Therefore, the septic field has to be moved closer to the front of the lot. Perhaps this is why the price was so low. I made up this example, but it could turn out to be all too real.

Can you see why it's important for a builder to take time to become familiar with well and septic systems? Chapter 3 provides a lot more information regarding what to look for when doing site inspections for well locations.

Bidding jobs

Bidding jobs can be tedious work. The job is lost when the bid is too high. Low bids result either in lost jobs because the bid is far too low or lost profits because the work is underestimated. It's not difficult to prepare a bid for a house that's to be served by well water. But, this type of bidding has some risks that don't affect builders who work with city water.

How would you bid a job for a house with a well? Are you going to seek quotes from well and pump installers, or are you simply going to factor in some amount for this work? You'd better get some firm prices. Guessing at the cost of a well system is a fast way to lose jobs and money.

The difference in cost between a shallow well and a deep well can be counted in thousands of dollars. That means, if you bid on the wrong type of well, it could cost you thousands of dollars. It is also conceivable that your bid price is going to be way too high if you bid on a deep well when your competitors bid on a shallow well. You must know the type of well you are bidding on before you commit to a price. Chapter 5 can help you understand the bidding process.

Water options

Water options are numerous. Wells are the most common source of private water, but they are not the only source. Springs serve a number of houses as a primary water source. Cisterns are sometimes used for water storage. Builders normally plan on using wells for rural homes, but even with that decision made, there is much more to consider. Dug wells, bored wells, drilled wells, driven wells, and other types of wells are all potential sources of drinking water. Knowing which type of well to use is important, and it can save you money. Sometimes the water requirements can best be met by some other type of water source. Chapters 7, 8, and 9 delve into this issue.

Minimizing your costs

Minimizing your costs to build a house can mean more money in your pocket. What you know about wells can make a difference in the price you have to pay for water. Chapter 14 deals with this subject, but let's talk about it here for a little while.

Assume that you are a spec builder. You're building a nice contemporary home on five acres out in the country. The building lot requires a well and septic system. Since this is a spec home, you're calling the shots on what is and isn't used in the construction. This means that it is up to you to pick a well and pump system.

You want to keep your well-related costs as low as possible. On the other hand, you can't afford to make a decision that might make the house harder to sell. A friend of yours has suggested a driven well. You're not familiar with this procedure, so you look into it. The process is cheap enough, but you're concerned that a driven well might not produce enough water on a regular basis to keep residents of the home satisfied. You're leaning towards a drilled well, because you've heard good things about them.

When you call four well installers to get prices on a drilled well, you are surprised that two of them recommend a shallow well. You're told that the water table in the area is high and consistent. Of the four installers you talked with, half of them suggested a shallow well. Now you are really confused. A friend has recommended a driven well. Two professionals have suggested a shallow well. You feel that a drilled well is the way to go. What are you going to do?

Being confused, you call the two well installers who gave you prices on a drilled well, as you had requested. You ask them their opinion. Wanting to make as much money as possible, both installers tell you all the advantages a drilled well has to offer. What they tell you is the truth, but that truth doesn't necessarily mean that you should install a drilled well. After a lot of research, you opt for a shallow well. This saves you a lot of money and gives you an adequate water source.

Is a drilled well better than a shallow well? Most professionals agree that it is. However, the added cost is not always justified. Under the right conditions, shallow wells can give very good service. Assuming that your well location can support a shallow well with a good recovery rate, there are not enough good reasons to justify spending all the extra money for a deep well. The savings allow you to price your spec house lower or to sell it for appraised value and pocket some extra money. Until you know the options that exist for wells, you can't hope to consistently keep your water costs down.

Problems

Builders know that problems with a house can pop up at almost any time and for almost any reason. Wells are no exception. Wells and their pump systems sometimes fail. If they malfunction while under your warranty period, you've got to take some action. This might be as simple as calling the subcontractors who installed the systems. However, you are the one who must deal with the customers. People can be difficult to deal with when they find themselves without water shortly after moving into a new house. It helps if you can provide some assistance or, at the least, some understanding.

Some problems that attack pump systems are simple to fix. They are so simple, in fact, that homeowners can often fix them with a little guidance. For example, a circuit-breaker might be tripped. If you know the symptoms of this problem, you might be able to talk the customer through a repair right on the phone. This quickly restores water service to your customer and saves your subcontractors an unwanted call-back.

While it is unlikely that you are going to venture into the field with an electrical meter and a box of plumbing tools, you should bone up on the problems associated with wells and pumps. It doesn't take long to get a cursory understanding of these systems, and the knowledge can come in very handy. Chapter 16 provides a host of troubleshooting tips. You can use this information to make repairs or simply to understand what your subcontractors are talking about when they report in to you.

You can lose a lot of money

You can lose a lot of money when you work with well systems. Depending upon your conditions, you could easily lose $1000 or more. This doesn't have to be a problem for you. There are ways to avoid losing money on well systems. In fact, you can use well systems to help you make more money.

Knowledge is the key to avoiding financial losses with well systems. This book offers plenty of information that can help you stay out of financial trouble created by mistakes with wells. Most of the wells that you deal with are going to be either bored wells or drilled wells. We are going to spend most of our time discussing these two types of wells. However, we are going to also look at driven wells and alternative water sources.

Before we leave this chapter, I'd like to give you a quick overview of pump systems. These systems are used with nearly every well that is made. Although pump systems don't carry as much financial risk as the well itself, pumps are very important.

Pump systems

When you install a well, you have to figure on installing a pump system. This type of system can be installed for less than $1000, but it can cost much more. Pump systems involve enough money to make them worth watching carefully.

Deciding on which type of pump system to use can get complicated. Not all pumps can work with all wells. For example, a one-pipe jet pump can't be used to extract water from a deep well. In the case of a deep well, you could use a submersible pump or a two-pipe jet pump. How do you decide which system is best? Research is the best way to find solid answers.

As you progress into later chapters, you are going to find plenty of advice on how to choose proper pump systems. Can you afford to rely on advice from the subcontractors who install pump systems for you? Most contractors are pretty honest. I'm both a builder and a plumbing contractor. If you called me as a plumber and asked my advice on pump systems, I'd give you straight answers. There is little reason for me to try to sell you one type of pump over another. Certainly, it is to my advantage to make sure you don't buy an inadequate system. Would I benefit from selling you one type of pump over another? It's possible, but I wouldn't risk my reputation by taking advantage of you. However, some pump dealers might not feel the same way that I do.

In order for you to know that installers are making solid suggestions to you, it's necessary for you to have a basic working understanding of pump systems. Pressure tanks are a good example. This is best illustrated with an example.

For our example, let's assume that you are a builder who is not accustomed to working with wells. Let's say that I'm a bad-guy plumber, looking to take advantage of you in any way that I can. You've asked me to give you an estimate for a pump system. The well where the pump is to be used is a deep one. Trying to protect yourself, you've done some homework and know that you want a submersible pump. As we talk, I do some informal probing to find out just how much you really know about well systems. It's my opinion that you are not much of an authority on the subject. This gives me a green light to set you up for a financial fall.

There are two ways for me to attempt to pick your pockets. I can pitch you on certain inferior products and offer you a price that I assume to be competitive with any other bidders who are supplying quality products. Or, I can try to sell you on a beefed-up system for a higher price. For the sake of our example, let me work you from the beefed-up angle.

If I sell you substandard materials for a high price and get caught, you can bring this back to me to create trouble. But, if I sell you a beefed-up system, you can't complain that I sold you equipment that was not rated for your job. The most you can say is that I sold you better equipment than was required. Is this so bad? Not nearly as bad as selling you something that is substandard. There could be a case made that I charged you too much, but this is a subjective complaint. All in all, I'm safer if I take advantage of you by providing better equipment rather than cheap or undersized stuff. Can I convince you to spend more for better equipment? Probably. I can sell you on the fear factor. If you don't know enough about pump systems, you can't tell if I'm blowing smoke or protecting you. This is my edge.

When we sit down to discuss pump options, you insist on a submersible pump. This is fine with me. As it turns out, you already know what brand of pump you want. But you don't know what size pump is needed. This is my first point of attack. I convince you to go with a three-quarter-horsepower pump instead of a half-horsepower pump. The smaller pump would work fine, but there is nothing wrong with using a larger pump, and I'll make more money by selling it to you.

The next step in my attack is the pressure tank. You have no idea of what size pressure tank is needed. To make the most money, I sell you on a large pressure tank. I do this by saying that while any pressure tank could be used, a large tank is much better and allows a pump to last longer. This is true, but a 40-gallon tank would do fine, and I've just sold you a 120-gallon tank.

How else can I get the price up? Well, I can pitch you on the use of gate or ball valves, instead of stop-and-waste valves. Maybe I'll lean on you to buy an in-line filter or some other type of water treatment condition. We could play this scenario out for several more pages, but it isn't necessary. By now, you have a good idea of what I'm trying to tell you. The only way to protect yourself is to learn all you can about wells and pump systems.

Wells and pump systems are not the only areas of home building that can put you at risk. Septic systems can present a lot of challenges. Indeed, septic systems are riskier than well systems. It is possible to find yourself in some major messes when you're involved with private sewage disposal. To expand on this, let's turn to the next chapter.

2

Septic systems can suck the profit right out of your job

You can lose a tremendous amount of money if you are not prepared to deal with septic systems. You could bid a job thinking that a simple septic system is going to cost around $4000 and then find out that a total cost of $10,000 is needed to install a suitable system. Can you afford to eat a $6000 loss? I surely can't.

Septic systems all perform similar functions, but not all systems are the same. In fact, many variations are possible with septic systems. Some are made mostly of gravel-and-pipe, along with concrete tanks. Other systems, which use chambers, are much more expensive. Occasionally, pumps are needed for septic systems. Pumps can really run the cost up. To be protected from financial losses, you need to have a good working knowledge of septic systems.

As a plumber, I might be able to take advantage of you on a pump system. At my best, or worse (depending upon your perspective), I can only take you for a few hundred dollars if I abuse my professional knowledge and sell you a pump system. Septic systems cost much more than pump systems, so it's possible to cheat you out of a lot more money.

If you deal with septic installers while you are ignorant of septic issues, you are taking some big chances. You owe it to yourself to become knowledgeable about septic installations. This book gives you the data that's needed to make you a savvy contractor.

Private sewage disposal

Private sewage disposal is often a factor in the construction of rural homes. Sewage disposal can be a volatile part of the job. Septic systems, when required, are a major part of a home's construction. A home can't be occupied unless it has a satisfactory septic system. As a builder, you must be aware of local requirements pertaining to septic systems.

When a person calls you to build a custom home that requires a septic system, you must be prepared to bid such a job. If you are a spec builder, you probably buy building lots either on your own or with the help of a real estate broker. When buying land, septic systems can be a deciding factor. How much do you know about soils studies and septic systems? If you don't know a lot about them, don't feel bad. Many builders don't know much about the inner workings of septic systems. However, if you don't know a good bit about septic systems, you should strive to learn about them. I assume you are doing that, since you are reading this book. Congratulations, you're on the right path to protecting yourself from financial losses.

Preliminary soils studies

Preliminary soils studies should always be done prior to purchasing a piece of land that requires a septic system. Most builders are aware of this, but some people new to the building business are not. Even experienced builders sometimes neglect the investigation of land before making a purchase. This is a big mistake. If you fail to take the proper precautions when buying land, you could wind up with a lot that is not buildable.

If you are bidding the construction of a new home, you should require an approved septic design before giving any prices. Without this design, you can't possibly bid a job properly.

Before purchasing land, spec builders should require at least a soils study. A soils design is safer and better, but a design is also more expensive than a study. But, wouldn't you rather pay a couple hundred dollars for a septic design to ensure that a lot is buildable, rather than gamble with the purchase of land at a cost of thousands of dollars? The answer seems simple enough, but some builders don't spend the time or money to properly check land. Hopefully, you're not one of these builders.

A septic design

Once you have an approved septic design, you are on much safer ground. Professional septic designs detail all aspects of a septic system. From a design, you can see what size septic tanks are required. The size and type of stone is specified. A piping layout is provided. Until you have an approved design, you can't accurately bid a septic system.

Septic subcontractors

Septic subcontractors often do more than just install septic systems. A majority of site contractors install septic systems. It's not uncommon to find that the contractor who clears, excavates, and grades your lot also installs the septic system.

Some septic contractors can guesstimate the cost of a septic system with amazing accuracy. But, they can't pinpoint the exact cost without an approved septic design. You are going to need a septic design before you can build a septic system, so you might as well get one early enough to save yourself the pain of buying an unbuildable lot. By the way, you need an approved septic design in order to get a septic permit, so there is really no way to avoid the cost of a design.

Types of septic systems

Several types of septic systems can be designed and installed. The most common type, and the least expensive, is a pipe-and-gravel system. The drain field on this type of system is made mostly with crushed stone and inexpensive perforated pipe. While these systems are not cheap, they are much less expensive than some other types.

Not all land can accommodate a pipe-and-gravel system. If the soil is not capable of absorbing the effluent from a septic tank quickly enough, a pipe-and-gravel system won't work. Only the best soil conditions are suitable for the least expensive septic systems. Luckily, pipe-and-gravel systems can be used at a whole lot of building spots.

As a spec builder, you can make more money if you find land where a simple septic system can be used. The difference in cost between a simple system and some of the more complex systems can exceed $6000. If you have a choice of two lots, both priced the same, and one requires a complex system while the other needs only a simple system, the septic difference can mean the equivalent of a nice discount.

Chamber systems

Chamber systems are more complicated septic systems. Chamber systems are used when soils can't absorb liquids quickly. The cost of a chamber system can be extremely high. If a job requires a chamber system, and you walk into it without knowing the chambers are needed, you are going to learn a costly lesson.

Pump systems

Pump systems are another expensive form of septic systems. If you have to combine a chamber system with a pump system, you are looking at big bucks. Can you imagine bidding a job for a $4000 pipe-and-gravel system only to find out that the job requires a $12,000 pump and chamber system? A mistake like that sends $8000 right down the drain, so to speak. You can't afford this type of problem.

A pump system is needed when the septic tank or field has to be installed at a higher elevation than the sewer of a home. While these systems aren't needed too often, they are not exactly rare. Every rural builder must be aware of pump systems.

As you move through chapters that follow this one, you are going to find a tremendous amount of information about various types of septic systems. By reading these chapters, you are going to be much better prepared for the real world of building houses with private sewage facilities.

You must be careful

You must be careful when working with septic systems. I'm not talking about the physical work, although safety on the job should always be a priority. The caution I refer to here has to do with money. When working with septic systems, you can lose big money in so many ways that you must take the time to protect yourself by building a paper shield. What's a paper shield? It's simply paperwork that puts all of your intentions and agreements in writing. If a job goes sour, this paperwork can save the day.

What could go wrong with something as simple as installing a septic system? Dozens of potential problems can affect a builder who is working with septic systems. Many of these problems are going to be covered in later chapters, but, I'll discuss some examples here.

Our first example

Let's say you are a builder who has dealt with septic systems in the past. While you don't build a lot of houses that require septic systems, you have had some experience with private disposal projects. This profile makes you a high-risk builder. If you had never dealt with a septic system, you would probably approach a job that required one with caution. A builder who works with septic systems on most jobs would have enough experience to avoid a lot of problems. You happen to fall into a dangerous category. You've installed enough septic systems to feel comfortable with them, but you really don't have extensive experience. Let's see what's going to happen.

You've recently acquired a piece of land for the purpose of building a spec house. A soils test and design has been completed. It appears that you can use a cost-effective pipe-and-gravel system. With everything looking good, you proceed with your project.

After obtaining your septic permit, you have the septic site cleared. Once the site is cleared, your septic contractor begins to dig. After just a short distance into the ground, the digger finds underground water, something you weren't expecting. A change in plans is in order. A mound system is required. Your building costs just shot way up. What went wrong?

When soils tests are done, the soil used for the tests is collected from random test holes or pits. Most people who do the tests use augers to get their test samples. Sometimes pits are dug with backhoes, post-hole diggers, or shovels. In all cases, the holes are generally dug in various locations within the boundary of a proposed septic site. Since the holes are made at random, it's possible for something like underground water to go unnoticed. The same could happen with a ridge of bedrock.

What can you do to protect yourself from this type of situation? It's not feasible to dig trenches through a septic area to expose underground obstacles prior to buying a piece of land. You could dig additional test holes. Digging the holes in a pattern, such as an "X" pattern, might reveal trouble before it's too late. Probing the ground with a probe rod could tip you to underground rock, but water would be more difficult to detect.

If you were building a home for a customer, you could make provisions in your contract to protect you from underground obstacles. Such a provision won't help a spec builder, but custom builders can write clauses into their contracts to limit their financial obligations when unexpected trouble is encountered.

I know that what we are talking about seems like a lot of extra work, but a lot of money is at stake. You don't have to go to extremes to prevent septic troubles, but failure to do so can be expensive. Use your own judgment, but I advise you to take proper precautions.

Another example

Let's look at another example of how septic systems can go bad. Assume that you have a septic design where everything looks good. You begin your septic work, and then it happens. Your subcontractor hits solid rock in the location where the septic tank is supposed to be buried. You've got three options. First, find a new place for the septic tank, which we are going to assume you can't do. Second, blast the bedrock to allow a deeper burial of a septic tank. Blasting could be done, but it's expensive. Plus there is some risk that it could collapse the well on this property. Based on the risks and expense, you rule out blasting. This leaves you with the third, an only remaining option—a pump system. Well, you're into an expensive problem.

What could you have done to prevent this problem? Some simple probing of the ground in the area of the proposed septic tank location would have revealed solid rock. Finding this rock early might have helped you prepare for your problem. Knowing that the rock exists is not a solution, but it could have made your work a little easier. At least you could have planned for better alternatives. If the rock was located soon enough, you might have avoided a pump system by making alterations to the foundation height of the house or to the exit location of the sewer. The point is this, preliminary research can't change the facts, but it can make dealing with them less destructive.

One more example

Let me give you one more example before moving on. In this scenario, you are a builder who is very experienced in the construction of homes where septic systems are installed. Your past experiences have taught you well. Based on your past history, you feel well qualified to bid jobs where septic systems are involved. Your comfort level might be a little too great.

In this example, you have been retained to build a large custom home. The house is going to contain five bedrooms. In your particular jurisdiction, the size of a septic system is determined by the number of bedrooms in a house. With a five-bedroom house, you are looking at a large septic system.

The house is on a pretty large lot. This is good, because the home is a large one. You get an approved septic permit and check it out carefully. Everything is in order, so you begin construction. This particular house has a three-bay, attached garage. Shortly after the roof is on, your customer comes to you and requests some extra work. Being a smart builder, you treat this request just like you would a new job. You figure the cost of the work and put all the details in writing. The customers review your proposal for a change order and signs it. Your performance seems satisfactory, but is it?

The extra work you have been requested to do involves creating a mother-in-law addition over the large garage. In doing this, you are adding two bedrooms over the garage. At this point, you have not realized that you have put yourself into a bad situation.

Work continues on the project. Everything is going well. Before drywall is hung, you get all of your rough-in inspections. However, a question is raised by the plumbing inspector. The approved plans on file with the code office show five bedrooms. According to the approved plans, there is no living space above the garage. But, you have signed a change order to provide finished space in the attic of the garage.

When the plumbing inspector sees rough-ins for a kitchen and bathroom over the garage, red flags spring up. The inspector contacts the code office by radio to confirm the fixture count on your plumber's permit. This leads to further investigation. It doesn't take long for the code office to figure out that something is wrong.

When you go by your job late that afternoon to see if it passed its plumbing inspection, you find a stop-work order on the job. This puzzles you, but the code office has closed for the day. When you contact the code officer the next day, you find that you are shut down. The job has been judged to be out of compliance. Why? Because you didn't take out additional permits to cover the work being done over the garage. You weren't trying to cheat the system, you simply forgot to apply for additional permits. But, you've got a big problem on your hands.

You have agreed, in a written contract with your customers, to supply additional living space over the garage. All of a sudden, you can't do this easily. It's more than a matter of paying for permits and a penalty fee. Due to the extra bathrooms, the existing septic system is no longer large enough. You have to upgrade the entire septic system to accommodate the additional bedrooms. Who's going to pay for this? I suspect that you are. Since you entered into a written agreement with your customers to give them extra space for a price, you have to stick to your promise.

Can you imagine how much it would cost to solve this problem? I can tell you that the expense would be significant. Could something like this really happen? It could. I know a builder who faced a very similar situation in my area. You can't afford to fall into traps like this one. You can avoid this type of problem if you remain alert and pay attention to detail.

What could you have done to stay out of harm's way in the above example? You should have known that adding bedrooms to a house would affect the size of a septic system. Being an experienced builder, you should have also known that permits would have to be amended to allow for additional work. By following proper procedures, you could have avoided the financial losses I have just described. All of the details of the job should have been worked out before you committed to a price with your customers. Trust me, avoiding a problem like this one is well worth whatever time you invest.

Firm prices

The only safe way to arrange for the installation of a septic system is to get firm prices from septic contractors. Offering to work with a contractor on a time-and-material basis is very risky. This is true in most types of trade work. But, the potential for high cost overruns with septic systems makes a time-and-material basis especially risky.

I worked as a consultant for a building company a few years ago. The company retained me to help them get started in the business of building homes. The first house I sold for the company to build required a septic system. A mistake was made by the company when contractors were brought onto this job. For example, the site contractor was allowed to work on a time-and-material basis. This contractor was also the septic contractor. There were some good reasons, in the minds of others, for allowing this contractor to work without a firm, contract price.

The septic installer had plenty of experience in site and septic work. However, the contractor had just recently opened his own business. He said that he didn't really know how to bid the job fairly, but that he needed the work and was willing to perform the needed services for a reasonable fee. The powers within the building company agreed to a time-and-material payment basis with the contractor. This fact served to hurt the contractor.

I was responsible for establishing this job. In doing so, I obtained site and septic prices from four local contractors. The prices varied substantially. Three out of the four contractors were well-established, reputable contractors. The bid prices seemed a little high to me, but all of the contractors had produced what I considered to be expensive bids. This encouraged me to seek other prices.

As I sought new contractors to bid the job, one of the carpenters suggested the contractor who wound up with the job. Based on recommendations from two trusted carpenters within our organization, a decision was made to award the job to the new kid on the block. This contractor would not have gotten the job without support from our carpenters.

Well, the site contractor did pretty good work. I've seen better, but his efforts were acceptable. What really went wrong was the cost of the work. By putting the contractor on a time-and-material basis, we basically gave him a blank check. I knew this was not a good idea, and I explained my fears to the people who had a vested interest in the building company. Based on their desires, I agreed to taking on the risk of a new contractor. It was a mistake.

The site and septic work on this job went on and on. Cost figures escalated. By the time the work was done, the total cost was higher than some of the bid prices from the other contractors. The building company would have been better off going with a fixed contract price. You live and learn, but the lessons can be costly. My advice to you is to only accept firm contract prices.

We've covered the basics on wells, pumps, and septic systems. Now it's time to get into the nitty-gritty work. We are about to move from a broad-brush approach into specific facts and instructions. You might wish to acquire a notebook and a pencil. It can be to your advantage to take a few notes as you move through the following chapters. At the very least, turn down the corners of any pages that you find particularly interesting. Now, let's move on.

3

On-site evaluations for water wells

On-site evaluations for water wells are often taken for granted by builders. I have known many builders who have submitted bids for work without ever seeing the building lot. This practice can be risky. One of the risks relates to the water well, when one is required.

A builder cannot bid a job competitively without knowing the type of well that is going to be used. If a bid goes in for a bored well when a drilled well is needed, the bidder loses money. If you try to compensate for not inspecting a site by basing your bid on the most expensive type of well, you stand a chance of unnecessarily bidding too high and losing the job. Failure to do a site inspection can be a very big mistake.

What can you really tell by walking a piece of land? It depends on your past experience and skill level. Some conditions point to obvious solutions. For example, when I built my last home, I was able to see bedrock sticking up in places. The type of humps in the land indicated rock close to the surface. A little scratching and digging proved that bedrock was right at the surface in some spots and not more than a couple of feet down in most locations. I automatically knew that a drilled well would be needed. When bedrock is present, drilling is the only sensible option.

You can't always see what's likely to be under the ground by looking at the surface. Knowledgeable builders want to know what they are going to get into when drilling wells and digging footings. Many experienced builders won't give a firm bid until customers or landowners provide them with soils studies. When such studies have not been done, some builders do their own. I'm one such builder.

It is not uncommon for me, or one of my people, to be out digging holes on potential building sites. A posthole digger can reveal a lot about the conditions that exist below the topsoil. Augers and probe rods can also provide some insight into what is likely to be encountered. A probe rod tells you if a lot of rock is present. But, to see the soils, you need a hole. An auger or posthole digger is the best way to get these samples. Augers are often easier to use. A power auger is even better, if you happen to have one.

When you create some test holes, you have a lot more information on which to base your bid decisions. A job that requires a well can only be bid in a few ways. You can guess what is going to be needed, but this is very risky. Digging test holes can give you a very good idea of what types of wells might be suitable. Interviewing well owners on surrounding property can provide a lot of data to help you with your decision. And, hiring soil-testing companies is a great, but expensive, way to find out what you are getting into. Another good way to protect yourself is to have a few well installers walk the land with you so that they can provide solid bids.

Since soils tests are required for land where a septic system is going to be used, the results from these tests can be reviewed to aid in the evaluation of a well type. This, however, is not always a safe way to proceed. Since septic locations are typically at least 100 feet from a well location, ground conditions might be very different. If you are going to rely on tests, the testing should be done at the proposed well site.

Since most builders are not well experts, it's wise to have well installers make site inspections in order to provide solid bids. If you have three to five bids from reputable well installers, you can feel secure in the fact that the bid you present a customer is going to be safe. This is the easy way out, but it's a good way to go.

You should get several firm quotes from well installers. It's risky to have only one bid. The well installer might be too busy to do your job when its time to start the work. It's possible that the installer could go out of business before you can request service from the company. As long as you have multiple bids from reputable installers, you should have very little to fear in terms of your well price.

Location

Choosing a well location is not always easy, but it's important. Many factors can influence the location of a well. The most obvious might be the location of a house. It is not common to place a well beneath a home, so most people choose a location outside of the foundation area of a home. Septic fields are another prime concern. Septic systems are required to be kept a certain distance from wells. The distance can vary from jurisdiction to jurisdiction. Also, the topography of the land can affect the distance. In addition, the well-drilling rig must be able to get to the site. These large trucks aren't as maneuverable as a pick-up truck. Picking a place for a well must be done with access in mind.

Appearance can be a factor in choosing a well location. Wells are not pretty, so they usually aren't welcome in locations where their presence is obvious. A drilled well is easier to hide than a bored well. The difference between a 6-inch well casing and a 3-foot well casing is considerable.

The location of a well also depends on where an expert believes water is going to be found. Few people know for sure where water can be found, but some people have a knack for being right more often than not. This brings us to the question of prospecting for water. Is it possible to predict where water can be found? A lot of people think so. Let's talk about this awhile.

Reading signs to find water

Have you ever heard of reading signs to find water? No, you won't find a billboard exclaiming "Water Is Here" along with an arrow pointing to the exact spot. However, an experienced eye can detect some natural signs that can reveal data on potential underground water.

Maps

Maps can give you a lot of guidance on where water might be found. Some regional authorities maintain records on wells already in existence. Reviewing this historical data can definitely help you pinpoint your well type and location. Unfortunately, there is never any guarantee that water is going to be where you think it is. A neighboring landowner might have a well that is 75 feet deep while your well turns out to be 150 deep. It is, however, likely that wells drilled in close proximity are going to average about the same depth. I've seen houses in subdivisions where one house has great well water and the next-door neighbor's water suffers from an unpleasant sulfur content. Fifty feet can make quite a difference in the depth, quantity, and quality of a well.

Topographical maps show land elevations. You can look at these maps and put many things into perspective. If there is a river or stream in the area, you can plot the position of your property in perspective to the surface water. Does this help you? Not necessarily.

My property has a good deal of river frontage. My well location is probably fifty feet, or so, above the river. Yet, my well is over 400 feet deep. I'm not sure why, but I think it has to do with the fact that my well is drilled into bedrock. Water doesn't run through solid rock very well.

Some maps are helpful, especially the ones that indicate depths and types of wells already in existence. Water does run through rock, but it needs cracks or some other form of access to get into and out of rock. All these factors make it difficult to predict what you are going to find when you attempt to install a well.

Plants

Plants can be very good indicators of water availability beneath the ground's surface. Trees and plants require water. The fact that plants and trees need water might not seem to provide much insight into underground water, but it can. Take cattails as an example. If you see cattails growing, you can count on finding water nearby. It's suggested that the depth of water in the earth can be predicted, to some extent, by the types of trees and plants in the area.

Supposedly, cane and reed indicate that water is within 10 feet of the ground's surface. Arrow weed means that water is within 20 feet of the surface. Many other types of plants and trees can indicate the presence of water. From a well-drilling point-of-view, I'm not sure about the accuracy of these predictions. I know that cattails and ferns indicate water is nearby, but I can't say that it's going to be potable water or how deep it is in the ground. I suspect, however, that there are some very good ways to predict water by studying the plants and trees.

Since I am not an expert in plants, trees, or finding water, I won't attempt to fill you with ideas about how to find water at a certain depth just because some particular plant grows in the area. I feel that with enough knowledge and research, a person can probably predict water depths with good accuracy in many cases. If water can be predicted with plants and trees, can it be found with a forked stick? Maybe so.

Dowsing

Dowsing is a subject that can cause a lot of controversy. Some people swear they can find water with nothing more than a forked stick. Can they do it? Sometimes they can. People have successfully located water with what is often called a witching stick. How hard is it to do this? There is a huge amount of underground water in the United States. Someone looking for water might be just as successful with a fireplace poker as with a twig from a willow tree. Since I have no first-hand experience with dowsing, I'm not in a position to make a judgment call on its merits.

I have known builders who hired dowsers to come to job sites and pick locations for wells. They probably always hit water. But then, so have I, and I've never used the services of a dowser. Rumor has it that the type of wood used for a divining rod (forked stick) is important. Favorite wood species include peach, willow, hazel and witch hazel. It's said that since these trees require a lot of water to grow, their wood is ideal for locating water.

I've studied the use of divining rods to a minor extent. Rather than forked sticks, people use metal rods to search for gold, water, and other underground treasures. As an amateur treasure hunter, I find it interesting to learn about new ways to locate hidden bounty. A television show about treasure hunting conducted a controlled test of self-proclaimed dowsing experts. The test was controlled over a man-made test bed. If my memory is correct, the dowsers did much better identifying water than they did finding gold. Did the testing prove anything? I don't know, but it was interesting to watch.

In my experience as a builder, I've never found a need to hire a professional dowser. You can try their services if you like, but I don't feel it's necessary. An experienced well installer who knows the area and has access to maps is probably just as good a bet, maybe better.

High-tech stuff
Some professional water hunters use high-tech equipment. I assume their sophisticated scientific gear works, but I don't know this to be true. If you were looking for an underground water source to serve a small community, it might be worthwhile to enlist the services of some of these professionals. But, for a single house, I wouldn't go to the trouble or expense. You are probably ahead if you hire a well driller, on a flat-rate basis, who can guarantee water.

Do a site inspection

It doesn't take very long to perform a site inspection. Usually an hour or so is all that's needed. Can you afford to spend thousands of dollars to save an hour or so? I can't. Site inspections are important. Whether it's you or your well installer that decides on the type of well that's needed, someone is going to assume a lot of responsibility. It's going to cost someone a lot of money if you planned to use a bored well and then found out that a 400-foot drilled well is needed. Customers expect you to guarantee them a firm price. You should expect the same from your well installer. While you might win more than you lose by playing the odds, the losses can be very expensive. Someone should definitely perform a site inspection before commitments are made on wells.

4

Site limitations for private sewage disposal systems

There are site limitations for private sewage disposal systems. Not all land is suitable for such systems. Land developers and builders should be able to spot land that is likely to give them trouble when it comes to a septic system. Experience does much to help in this area. After several years of buying and developing land, a person begins to recognize a good site.

Little signs can give you hints about the quality of land. For example, bulges and occasional glimpses of rock on the land's surface could mean that bedrock is close to the surface. Bedrock can interfere with a private sewage disposal system.

When I was developing land and building homes in Virginia, septic systems were a common part of my job. In all the jobs I did in Virginia that had septic systems, I never encountered bedrock. Most of the soil conditions were very good for leach fields. This has not been the case since I moved to Maine.

Land in Maine, where I now work, is rocky. Most of the rock is underground, but not by much. Bedrock, or ledge as it's called in Maine, can be within inches of the topsoil. It took me awhile to get used to this fact. It affects the installation of septic systems and foundations. Quoting a price for a job without knowing that ledge is present can be a financial disaster. If you have to blast bedrock to get in a full foundation, and you haven't planned for it, your profits can be destroyed.

Rock is not the only risk when it comes to septic systems. Some ground just won't perk. If this is the case, the land might be deemed unbuildable. Buying land for development and then discovering that it has no acceptable septic sites can really ruin your day. Experienced land buyers put clauses in their purchase agreements to protect them from this type of risk. A typical land agreement contains a contingency clause that gives the buyer a chance to have the soil tested before an absolute commitment is made to buy the land. If the tests prove favorable, the deal goes through. When soils studies turn up problems, the contract could be voided or some adjustment might be made to the sales price.

The size of a building lot sometimes affects whether or not it can be approved for a septic system. Many houses that require a septic system also require a water well for drinking water. For obvious reasons, septic systems must not be installed too close to a water well. The minimum distance between the two is normally 100 feet, but it sometimes is more. I have seen exceptions that allowed less distance, but not many.

Since moving to Maine, I've seen some building practices that I never experienced in Virginia. For one thing, I can't imagine anyone getting a building permit and putting in a foundation before their well and septic locations have been chosen and approved. Yet, I know of a recent case where this happened. Let me tell you about it, so the same problem won't come up with your building business.

A man in Maine bought a piece of land and hired a builder to build his house. The builder had his site contractor clear the lot and install the septic system. It was passed without any problem by the local code officer. Then the builder put in the footings and full basement walls. It was about this time that a problem was discovered.

The house needed a well. But, because of the septic system location, there was no room on the lot to install a well that would comply with code requirements. It is not that the lot was too small for both a well and septic system. It was the location of the septic system that caused the problem. Basically, the builder created an unbuildable lot from one that had been buildable. I would not have wanted to be in that builder's shoes.

In an effort to salvage his situation, the builder went to adjoining landowners. He asked for an easement so that the well for his house could be drilled on their property. It took some doing, but I believe an agreement was worked out. I don't know what would have happened if the neighbors refused to cooperate. I suspect the builder would have lost a costly lawsuit.

The builder we just discussed made some very serious errors in judgment. He was lucky that the circumstances didn't turn too ugly. When building a house, you are at risk any time that you must rely on a private septic system. And, it's not only houses that can be at risk. Some commercial buildings depend upon private waste disposal systems. You must know what to look out for in order to protect yourself.

Let me tell you about a builder in Virginia who made a big mistake when bidding on a job. This builder went out to a building lot with the lot owner. After walking the property, the builder accepted a set of plans and went back to his office to work up a price for a new house.

After completing a quote, the builder was awarded a contract to build the house. It was not until the permit process started that the builder found out that the municipal sewer, which served many houses in the area, didn't extend to the building lot where he was going to build. There was a public water supply, but no sewer. This left the builder with two options. He could either pay to have the street cut and the sewer extended, or he could pay for a septic system.

When the builder priced this project, he planned to pay a tap fee to connect to a city sewer. As I remember, the tap fee was around $3000. At any rate, the only money he had budgeted was the amount figured for the tap fee. It would cost much more to have the sewer extended. Installing a septic system would be difficult, due to the size and shape of the lot. A deal was worked out with the lot owner and the sewer was extended. I don't know what type of financial adjustments were made between the builder and his customer, but I expect the builder made considerably less money than he had originally planned.

As they appear

Things are not always as they appear. In the case of our last example, the builder assumed that public water and a sewer were available for the building lot. Unfortunately, the sewer connection was not readily available. This situation is not as rare as you might think.

I've run into a number of building lots that have municipal hookups for water, but nothing available for private waste disposal systems. With my contingency clauses and inspections, I've never been put in a bind from this type of problem. But, a buyer or builder who doesn't research the options that are available could get in big trouble very quickly.

Some people assume that a lot is going to be suitable for a septic system just because lots on either side of it use septic systems. This is not always the case. I've seen property where, out of five acres of land, there might be only one or two sites suitable for a standard septic system. This situation is not uncommon, so watch out. Make sure that you have an approved septic location established before you make any firm commitments to buy or build.

Grade

The grade of a building lot can affect the type of septic system that can be installed. If the grade does not allow for a gravity system, the price goes up considerably. Pump stations can be used, but they are expensive. If you're building houses on spec, you might have trouble selling one that relies on a pump station. People don't like the thought of replacing pumps at some time in the future. And, many people are afraid the pumps are going to fail, leaving them without sanitary conditions until the pump can be replaced.

The naked eye is a natural wonder, but it can't always accurately detect the slope of a piece of land. Looks can be deceiving. Unless the land leaves no room for doubt, it's best to check the elevation with a transit. You can't afford to bid on a regular septic system and then wind up having to install pump systems.

A safe way

There is a safe way to work with land when a septic system is required. You can go by reports provided by experts. Even this isn't foolproof, but it's as good as it gets. If you have a soils engineer design a septic system, you can be pretty sure that it is going to work pretty much as drawn. Since every jurisdiction I know requires a septic design before issuing a building permit, it makes sense to go ahead and get the design completed early in the game.

I've dealt with two types of situations when getting septic designs. In Virginia, the designs were drawn by a county official. The cost, if there was one, was very minimal. Let me tell you how this process worked.

A representative of the county would come out and bore test pits with an auger. This was done by hand, so there was no need to clear the land for heavy equipment. After digging the pits, a perk test was done. A few days later, a finished design was provided by the county. It didn't take long or cost much to have a detailed plan created. The plan was official, so it could be counted on for permit acquisition.

Things are done differently where I work in Maine. Here, soils studies are done by independent engineers. The cost for these studies is normally around $250. As a builder, I hire an engineer to test the soils and draw a septic design. I can then use that design to obtain a septic permit from the code enforcement office. This process is more expensive than the one in Virginia, but in the scheme of things, it is still a bargain. I'd rather pay $250 to avoid walking into a problem. It beats spending thousands of dollars trying to fix a mistake.

If you are buying land, make sure that your purchase agreement provides a contingency allowing you to have the land approved for a septic system before you are committed to going through with the sale. It is wise to specify in your contingency what type or types of septic designs are acceptable. Some types of systems cost much more than others. If your contract merely states that the land must be suitable for a septic system, you would have to complete the sale regardless of the system cost, as long as one can be installed.

When bidding jobs, your quote should make clear what your limits are in regards to site conditions. Specify the type of septic system your quote is based on, and provide language that protects you if some other type of system becomes required. Don't take any chances when dealing with septic systems, because the money you lose can be significant.

A standard system

A standard septic system is built with a gravity flow. Its drain field, or leach field as it is often called, is made up of perforated pipe and crushed stone. This is the least expensive septic system to install. Unless the soil does not perk well, you normally can use a standard system.

Chamber systems

Chamber systems are much more expensive than pipe-and-gravel systems. Chapter 11 describes these systems in detail, but let me give you a brief description of them here, so you can see how they can affect your site selections.

Chamber systems are used when soil doesn't perk well. If the soil can perk, but not perk well, a chamber system might be your only choice. How do these systems work? Basically, the chambers hold effluent from the septic tank until it can be absorbed by the ground. Unlike a perforated pipe, which releases effluent quickly, the chamber controls the flow of effluent at a rate acceptable to the soil conditions. Where a pipe-and-gravel system might flood an area with effluent, a chamber system can distribute the liquid more slowly and under controlled circumstances.

The cost of a chamber system can easily be twice that of a pipe-and-gravel system. In my area, it's not unusual for a chamber system to cost up to $10,000—that's a lot of money for a septic system. And, if you happen to bid a job based on a gravel system, at say $4000, the extra $6000 spent on chambers is going to deflate your profit very quickly.

Pump stations

Pump stations can be another big expense in some septic systems. The cost of the pump station, the pump and its control, and the added labor for installation can add thousands of dollars to the cost of a standard septic system. If you were unlucky enough to get stuck with a pumped chamber system when you had planned on a gravity gravel system, you could lose much of your building profit all at once.

Trees

Trees are another factor to consider when doing a site inspection for a septic system. Tree roots and drain fields don't mix very well. Any trees in the area of the septic system must be removed. Even trees near the edges of the area should be removed. How much open space should exist between a septic system and trees? It depends on the types of trees that are growing in the area. Some trees have roots that reach out much farther than others. You can get a good idea of how far the roots extend by looking at the branches on the tree. The spread of the branches is often similar to the spread of the roots. If there is any doubt, the professional who draws your septic design can tell you which trees to leave standing.

Burying a septic tank

Burying a septic tank requires a fairly deep hole. Even if you are using a low-profile tank, the depth requirement can be several feet. If you work in an area where bedrock is present, as I do, you must be cautious. You could run into a situation where rock prevents you from burying a septic tank. This was almost the case on my personal home.

When I built my most recent home, the land I chose consisted of a lot of bedrock. When digging footings for the house, my site contractor hit solid rock in less than two feet of excavation. I knew we would hit rock at a shallow depth, so this didn't alarm me. It did, however, make it difficult to install the water line from my well at a depth that would prevent it from freezing. In fact, I had to use an in-line heat tape to protect the water service. The rock also created some concern on how we would bury my septic tank.

When I bought the land, I probed it with a metal rod to determine how far below the surface I would encounter ledge. Numerous probe sites showed that 18 inches was the most I would be able to dig in many areas. However, in some areas, the bedrock was deeper. With the use of my probe, I was able to plot a location for the foundation, well, and septic system.

My site work turned out pretty much as expected. Since I had probed the soil, there were no big surprises. By using the natural slope of the land and some fill, I was able to bury a low-profile septic tank without any real trouble. However, if the land had not consisted of a natural slope, I might not have been able to get my septic tank buried without blasting the bedrock. It might have been necessary for me to build some type of mound system to overcome the problems associated with rock. This would have been very expensive. My preliminary site evaluation proved very helpful in setting and staying within a budget.

A stiff, small-diameter steel rod with a sharp point is very helpful when probing septic locations. The rod can be used before a system is installed to establish any underground obstacles. But, make sure that the area you are probing does not contain any underground utilities. Jamming your probe rod through buried electrical wires can be a shocking experience.

After a septic system is installed, a probe rod can be used to locate the septic tank if you ever have to find it again. Someone should pinpoint the tank location and record the information for future reference. The property owner needs access to the tank openings in order to have the system pumped out periodically. In any event, a good probe rod comes in handy.

Underground water

Underground water can present problems when installing a septic system. This problem is usually detected when test pits are dug for perk tests. However, it is possible for the path of the water to evade detection until full-scale excavation starts. For this reason, you should have some type of language in your contracts to indemnify you against underground obstacles, such as water.

I've never had a problem with water when installing a septic system, but I have had it get in the way while remodeling a basement. My helper and I broke up a concrete floor in a basement once and found a fast-flowing stream just below it. The water was so abundant that it had washed out the crushed stone used under the concrete. If this much water can run under someone's basement floor, it could certainly pose a problem for a septic system.

Driveways and parking areas

When you assess a lot for a septic system, consider the placement of driveways and parking areas. Even though a septic system is below ground, it is not wise to drive vehicles over the system. The weight and movement could damage the drain field to a point where it would require replacement. You don't need this type of warranty work, so make sure that all vehicular traffic avoids the septic system.

Erosion

Erosion can be a problem with some building lots and land. If you install a septic field on the side of a hill, you must make sure that the soil covering the field remains in place. This can be done by planting grass or some other ground cover. When checking out a piece of land, you need to take the erosion factor into consideration. The cost of preventing a wash-out over the septic system can add a significant amount of expense to your job.

Set-backs

Set-backs are something else that you should check on before committing to a septic design. Many localities require that all improvements made on a piece of land must be kept a minimum distance from the property lines. A typical set-back for a side property line is 15 feet, but this is not always the case. Where I live, there are no set-backs. But, I've seen set-back requirements that were more than 15 feet. This can become a big factor in the installation of a septic system. Let me give you an example.

Let's say that you are buying a piece of land to develop into building lots. You've had your perk tests done and your septic designs drawn. On paper, there doesn't seem to be any problem. All of the wells and septic systems fit on their individual lots according to the plot plan. But, when you go to get your permits, you find that the zoning requirements establish side set-backs of 20 feet. Based on this, some of your septic systems encroach on the set-back zone. This, of course, makes them unacceptable. You might be able to obtain a variance that would allow the drain fields to run into the set-back zone, but not necessarily. If you can't bend the rules, you can't install the septic systems. Do your homework, and make sure that what you have sketched on paper can actually be installed.

Read the fine print

When you are looking at a septic design, it's important to read the fine print. A design gives you all the specifications needed to bid a job. You should read the design carefully to determine exactly what types of materials are specified. It is a good idea to take the design with you when you do an on-site inspection. A long tape measure should also be considered standard equipment in your field inspections.

When you arrive at the building site, use the septic design to lay out the system. Once you have the septic site staked out, you can see more clearly any obstacles, such as trees, that might be in your way. When you stake out the field, you don't normally have to be precise. The idea here is to see approximately what you are dealing with in terms of a work area. How difficult is it going to be to get equipment to the site? This is something you might not think about, but you should.

A few years ago, I oversaw the construction of a house that had a leach field high on top of a hill. A pump station was installed behind the house to pump effluent up to the field. The topography made it difficult to get crushed stone to the site. Trucks couldn't back up to the work area and dump their loads. Instead, the gravel had to be ferried up the hill in the bucket of a backhoe. This procedure didn't make the job impossible, but it did make it much more expensive. As a bidding contractor, you have to keep your eyes open for such problems.

By studying the septic design, you can determine the length of the drain lines. Also, there are details that show the depth of the trenches. With this information, you can use your probe rod to take random tests for underground obstructions.

Is the work area large enough to accommodate the piles of dirt and gravel that accumulate as the work progresses? If limited space forces you to work the field in sections, it's going to take longer to construct the system. Whenever there is an increase in time, there is usually an increase in cost.

If trees must be removed, what are you going to do with the stumps? Getting rid of tree stumps has become a problem in the area I work. At one time, it was common practice to bury the stumps, but this is no longer the case. More often than not, stumps are hauled away to be ground up. The hauling and disposal fees for stumps can add a lot to the cost of a septic system. Keep this in mind if you are responsible for clearing the work area and disposing of unwanted debris.

Pay attention to all the details on the septic design. Everything needed to price a septic system should be on the design. Once you combine the design with a physical inspection, you should be in a good position to work up an accurate price.

Site visits

Site visits should be considered a mandatory step when bidding jobs that require private waste disposal systems. You can tell a lot from a septic design, but you can't formulate a safe bid until you walk the land where the system is to be installed.

Many contractors try to use averages when figuring jobs. For example, they include so much per square foot for framing costs and so much per square foot for siding work. They might use a per-fixture price to estimate plumbing costs. This method can work. I've estimated jobs based on a square-foot factor or fixture factor, but you can never be very sure about this method of pricing. It's even more risky to use averages when a septic system is involved.

Unless you do a full-blown take-off and work-up on a job, you can't be sure of your prices. Even when you go to the trouble of figuring every little cost, there is always some risk of error. Considering how much money could be at stake with a septic system, it would be very dangerous to simply plug a generic figure into a bid proposal.

It's important to get quotes from contractors who are in the business of installing septic systems. You should get the quotes before you offer prices to customers. Builders and general contractors do not always have enough experience in each trade to produce accurate estimates on their own. This is why wise builders get firm prices from subcontractors before presenting any price to their customers.

Assuming that you are going to be acting as a general contractor, you should take your septic contractor out to the job site for a first-hand look at the lay of the land. Unless you insist that your subcontractor visit the work area, you can't be sure that you are going to get an accurate price. Most septic contractors won't offer you a price until completing a field inspection. Contractors who are willing to give you phone quotes are playing the odds. As long as you know they are going to stand behind their prices, you can go along with them. But, only you can decide if you are willing to take the risk of doing without a physical inspection of the work site. Personally, I wouldn't trust a quote from anyone who hadn't first looked at the site.

5

Protect yourself during the bidding phase

To be a successful contractor, you must protect yourself during the bidding phase. This can be done in many ways. One of the most important methods requires you to put everything in writing. Verbal statements rarely hold water in court. You need written documentation in order to have solid footing in a legal battle.

You've seen some of the many risks associated with wells and septic systems. When you assume responsibility for installing these systems, you are putting yourself at risk. This responsibility, however, is part of your job. But, the risk factor can be kept in check with a little thought and preparation on your part.

What type of proposal do you use? Is it one of those generic, fill-in-the-blank forms that you order from a catalog? This type of proposal and contract is used by a lot of contractors. Just because the forms are used in large numbers doesn't mean that they are good. You should have your attorney draft forms for you to use.

Even the best forms in the world are not much good unless you use them. How many times have you given customers a quote over the phone? Do you have any record of your conversation? I doubt it, and even if you did, it probably wouldn't be enough to save you in a court battle. Your bids should be given in writing. If you have to give a price over the phone, follow it up with a written proposal.

Confusion is one of the biggest reasons why contractors get into arguments with customers. If your intentions are spelled out in black and white, there is much less room for confusion to occur. Until you and your customer understand each other's plans completely, there is unneeded risk of a conflict.

As a builder, you have to take responsibility for a lot of various building trades. You are probably more familiar with some of the trades than you are with others. This is only natural. Regardless of how much you know, or think you know, about a trade other than your own, you can't afford to make representations without documentation. And, you might not be able to create solid documentation without some help from experts in the fields of work that you are dealing with at the time.

Look at the last contract you signed with your well driller. Pay particular attention to the disclaimers. Now, do the same thing with the last contract you signed with your septic installer. Compare those contracts with the proposals you presented to your customers for those jobs. Does your bid package detail the same disclaimers found in the proposals from your subcontractors? If it doesn't, you are assuming too much risk.

Many contractors take a casual approach when bidding jobs. I've seen bids for major work come into my office with only a couple of paragraphs describing the work and terms being offered. This not only makes for a poor proposal in terms of sales appeal, it leaves a lot of room for confusion and confrontation.

Your formal quote to a customer is a serious communication. If you leave details out of your quote that are later put into a contract, you might run into resistance when you ask your customer to sign your contract. I believe the quote should provide complete details of every offer you are making. If there are exclusions, they should be spelled out in the quote. Any substitution of materials should also be put in clear language. Your quote is your sales presentation, so it should be accurate and enticing.

As a general contractor or builder, I'm sure that you've received numerous quotes and estimates from subcontractors. You must have seen some pretty skimpy ones from time to time. My experience has shown that a majority of bids are poorly prepared. Customers take accuracy, neatness, and thoroughness into consideration when trying to decide who gets their contract. But, getting the job is not your only concern.

Once you have won a bid, you want to make money and keep your customer happy. Detailed quotes and contracts are a good step in reaching your goal. If pertinent details are omitted from either a quote or a contract, you might very well have a disgruntled customer on your hands. It's possible that you might even be dragged into court. I can give you an example of this to consider.

One of my carpentry subcontractors was telling me a story about a house he recently built for a customer. As part of the contractor's contract, he was supposed to have his site subcontractor provide a certain amount of loam for a job. During the construction process, the site contractor and the owner of the property had a discussion. They agreed jointly to use sand in place of loam for much of the fill work, because sand would be cheaper. A written change order was not used to document this new agreement. The site contractor performed the services as agreed to by the homeowner.

After the house was built, the homeowner refused to pay the subcontractor for some of the work that had been done. One of the phases being contested was the site work. The owner wouldn't pay up, so the site contractor took him to court. As the general contractor, my friend was also required to make an appearance in court.

The court hearing didn't last very long. After some verbal exchanges and accusations, the judge referred to the only written agreement between the parties. The contract called for a certain amount of loam. When the judge asked if the described amount of loam had been installed, the site contractor had to admit that it had not. He tried to qualify his answer by telling the court about the verbal agreement to substitute sand for a portion of the loam, but the judge wouldn't hear it. Based on the written agreement, the court ruled in favor of the homeowner.

If everything I was told was true, and I believe it was, the site contractor got a raw deal. I can't fault the judge for going by the only real evidence available, but the homeowner took advantage of the site contractor. If you want to avoid this type of problem, make sure that all of your agreements are in writing. This pertains to all aspects of your deal, including any exclusions.

Disclosure

Every well proposal I've seen has included some form of disclosure. One statement commonly found in well proposals usually reads something like this: "the driller will not guarantee the quantity or quality of the water produced by a well." This is only one simple sentence in the quote or contract, but it can have a huge impact. Let's talk about this for a few minutes.

Assume that you are building a new house for a young couple. The house is in a rural location and requires both a well and septic system. As the builder, you are responsible for all aspects of the job, including the well. Further assume that your agreement with the well driller does contain some version of the standard disclosure about quantity and quality. Now let's say that the well driller hits water and everything is fine until about six months later when the dry season rolls around.

One day the homeowner calls and complains that the house you built has no water. After troubleshooting the situation, you discover that the well has run dry. As the builder, you have to warranty the house for one year. After only six months, the well has dried up. What's going to happen?

The homeowner is probably going to expect you to dig or drill a deeper well. If you ask your well driller to do the job as warranty work, you are probably going to be told to forget it. The driller can simply produce the contract you signed, the one with the innocent little sentence in it, and prove that the quantity of water was not guaranteed.

Where does this leave you? Sort of between a rock and a hard spot. You've got to give the customer a better well, and your well driller doesn't have to do any of the work as warranty work. Under the circumstances, you are going to have to pay for the new well out of your own pocket.

You could probably avoid this situation by putting the well contractor's disclosure into your contract with the homeowner. If the homeowner was getting the same deal as you were, and the facts were set down in writing and signed, you'd be in a much better position to avoid paying for a new well.

The use of disclosures and liability waivers can go a long way towards protecting you and your business. If you have specific terms written clearly into any agreement you make with a customer, you are much less likely to get into arguments with your customers. This is not only good protection, but good public relations as well.

Danger spots

What are some of the danger spots that should make you extra cautious when bidding jobs with wells and septic systems? Many circumstances can put you at risk. Some of them, but not all, can be avoided. With the use of disclosures, exclusions, and liability waivers, you can put up a pretty good shield around you and your business. To expand on this, let's look at some of the specific areas of concern. Let's start with septic systems.

Septic systems

Septic systems are fairly simple to install. Yet, complications can get in the way. There can be a great deal of risk when bidding jobs requiring private sewage systems. To help protect yourself and your company, you need to address some of these key issues in your proposals and contracts.

Septic permits

Septic permits are usually required before starting the installation of a septic system. Normally, this is not a problem, but it could become one. When bidding a job, you should make your price and work subject to the issuance of a septic permit. Then, if for any reason, a permit cannot be obtained, you have an out.

Having a request for a septic permit denied is not common, but it can occur. It could be due to poor soil conditions, but there is another reason why the request might be turned down. In some areas, as public sewer lines are extended, they sometimes end up located near homes with septic systems. Assuming a building lot is large enough to accommodate a septic system, a property owner might prefer having a private system. This eliminates the tap fee charged by the sewer authorities to connect to the public sewer. It also does away with monthly usage fees for the public sewer. However, the jurisdiction issuing permits might not allow the installation of a private system when a public sewer is available. This is a common practice. You could walk into a nightmare if your quotes and contracts don't cover this possibility.

A specific design

When you write up a quote or contract to install a septic system, you should name a specific type of design. If your quote is based on a pipe-and-gravel system, say so in your quote. Reference the septic design drawing that you used to calculate your pricing. If for some reason the specified design isn't approved, you're at much less risk. An open clause that simply states that you are going to install a septic system for a certain price is much more dangerous. The more specific you are in your language, the less likely you are to have a problem.

Clearing

An area must often be cleared in order to install a septic field. Who's responsible for this work? If your proposal says you are to install a septic system and does not exclude clearing, it could (and probably should) be interpreted that you are going to take care of all aspects of the job, including clearing.

If you are going to clear land for a septic system, detail in your proposal the work that is to be done. How large an area is to be cleared? The answer to this question should be a part of your proposal. Are you going to remove stumps, rocks, and other left-over debris? If not, exclude this work. Otherwise, stipulate what you plan to do with the debris. Make all of your work description very clear.

Access

Access to a septic site can be a problem. You might have to cut down trees to get trucks and equipment to the site. If your customer is not aware that trees outside of the septic area are going to be removed, you might have an angry customer on your hands. Indeed, you could be in big trouble if the customer comes out to inspect your progress and then is shocked to discover you removed some favorite trees without giving any notice. If an access path must be made, be specific about where it is going and what is to be done.

I've had occasions when the only practical way to get equipment to a septic site was to cross over the land of adjoining owners. Don't put equipment on adjoining property until you have written permission. If your customer tells you that the neighbors won't mind if you use their driveway or land for access, don't believe it until you have it in writing.

Rock

Rock can make the normal installation of a septic system impossible. You should address this issue, as well as other underground obstacles, in your proposal and contract. I'll leave it up to you and your lawyer to work out the exact wording, but make sure that you have some type of protection in the event that unseen obstacles prevent you from doing the work you are proposing at the price you are quoting.

Materials

The availability and price of materials is another issue that you might want to cover in your protective paperwork. I don't believe this is a big issue with septic systems, but it wouldn't hurt to have some type of clause in your agreements to deal with rising prices and materials that are not readily available.

The sewer

Who is going to run the sewer from the house to the septic tank? Septic installers often run the sewer to within 5 feet of the house foundation. From there, a plumber takes over. This, however, is not always the case. Sometimes plumbers do the sewers. If you are bidding all phases of a job, this is not such a big deal. However, if you are bidding the septic work and not the plumbing work, a question as to who is responsible for the sewer could become an issue. Digging the trench for a sewer can take some time. Also, the labor and materials needed to install the pipe must be considered. Keep the sewer installation in mind when you are bidding your next job.

Landscaping

Are you going to landscape the septic area? Who does the finish grade work? Who is going to seed and straw the area? These questions should be answered in your proposal and contract. Even the type of dirt used to cover the excavated site should be detailed.

Perimeter trees

Perimeter trees can interfere with a septic system. If trees are left standing too close to a drain field, their roots might invade the field. If you feel that perimeter trees should be taken down, stress your point in your proposal. Owners who prohibit you from following your instincts on this issue should be asked to sign a liability waiver that releases you from any damage done by the perimeter trees.

Wells

Wells, like septic systems, can present bidding contractors with some problems. Certain clauses often found in well-drilling contracts can set a builder up for trouble. If you don't provide some protection for yourself in the contracts, you could lose thousands of dollars. A lot of houses depend on wells for potable water, so you better be prepared to bid on those jobs. Let me give you a few pointers on what to look out for.

By the job

Are you paying your well driller by the job or by the foot? Both payment methods are usually available. Depending upon circumstances, and the payment method you choose, there can be dramatic differences in the overall cost of producing a well. Since this is an important subject, let me give you a little background on how I worked with wells in the past.

The well drillers I work with always give me two payment options. I can pay so much for each foot of well depth, or I can pay a flat rate that doesn't change no matter how deep the well. The flat-rate price also guarantees me water, so if a driller hits a dry hole, I'm not paying to have the second well sunk. Under these conditions, what way would you go?

The flat-rate price for a well is usually pretty steep. Well drillers generally look at historical data for the existing wells in the area.

They calculate a worst-case scenario and base their prices on that information. The per-foot price often works out to be cheaper, sometimes as much as $1000, but there is additional risk involved for the person paying the bill.

In the past, I've always gambled on my wells. Even though the guaranteed deals are safe, I wanted to try and save the extra money if I could. Like the well drillers, I researched the area well depths. Sometimes I would talk to owners of adjoining properties to find out the depth of their wells. This gave me a good idea, but no guarantee, of how deep my wells would have to be. Using this strategy, I've saved lots of money over the years. I could have gotten caught on some, and had to pay for extra-deep wells or even a dry hole, but I never have.

When I built my most recent personal home, I opted for a guaranteed price. It was the first time in all of my years as builder that I ever went with a flat-rate fee. And, am I ever glad that I did.

Most wells in my area are between 250 and 300 feet deep. If I were estimating my costs on a per-foot basis, I would have used the 300-foot figure. I have river frontage on my land, and I thought the water vein might be even closer. My neighbors live about one-half a mile away, so their wells are not a great barometer to use, but they are better than nothing. I figured my well would not be deeper than 300 feet and might be as shallow as 200 feet. Yet, something gnawed at me about this well. For whatever reason, I decided to go with a fixed price. My well wound up being 404 feet deep. Even the driller was shocked. In this case, the driller lost and I won. If I'd gone on a per-foot price, the cost would have been well above my budgeted amount.

Which pricing method should you use? If you want to be safe, go with the guaranteed price. It's up to you to decide which way to go, but you should put something in your proposal to identify the conditions you are giving your customer. You might say that the cost of the well is going to be a specific amount and that the amount is guaranteed. The price might scare your customers, but at least they can rest easy knowing there won't be any surprises.

As another option, you could give the customer a price based on so much per foot for an estimated well depth, say 250 feet. If the actual depth turns out to be less, you credit the difference to the customer. Should the depth be more, the customer pays the extra cost. This approach might frighten your customers more than the higher fixed price.

As a matter of practice, I give my customers the same option that the well drillers give me. I let them choose between the per-foot and the fixed-price method. Once they have made a decision, I memorialize it in our agreement. This method has always worked well for me.

I've been lucky with my wells. Few have run deeper than expected, and I've never run up against a dry hole. Other contractors I know have not been so fortunate. It's no bargain when you pay for two or three wells in order to get one, so be careful on this issue.

Drilled or dug

Is the well you are providing to be drilled or dug? You could simply put in your agreement that you are providing a well. Most customers wouldn't question this approach. However, after the well is installed, you could get some grief if you installed a dug well and the customer thought a drilled well was being installed. You should clearly identify the type of well your price includes.

Dug wells are cheaper than drilled wells, but they sometimes run dry. Drilled wells rarely run out of water because of their extreme depth. If your customer is looking to cut corners and the local ground conditions allow it, a dug well and a jet pump are the least expensive way to go while still maintaining a viable hope for a steady water supply.

Driven wells, in my opinion, should not be considered a worthwhile option for full-time homes. These little wells can produce good water, but the quantity is limited. While a driven well makes sense for a week-end cottage or fishing camp, I don't feel it is suitable for the demands of full-time use.

Drilled wells are the most expensive to install, but they are also the most dependable. A submersible pump and a drilled well are the way to go in my opinion. In the several houses I've owned, I've had experience with dug wells and drilled wells. In my opinion, the drilled well can't be beat.

Well preference should be left to the customer's discretion. However, in your paperwork, you need to make it clear what type of well is wanted, what your prices are based on, and how problems that might arise are going to be handled.

Quantity

Will your customer be guaranteed a certain quantity of water from a well? Your well driller probably won't make this commitment, so you shouldn't either. Every driller's proposal I've seen excludes any assurance of water quantity. This is a touchy subject, but it is one you should deal with before a problem pops up. If your customer complains about a lack of water quantity, you are unlikely to have anywhere to turn. Think about this.

How can you get your customer to accept the fact that you can't guarantee a quantity of water. There is a method that I've used for years that has never let me down. Get several quotes from well drillers. If you're not afraid to show your customers the prices that you are paying, show them the actual quotes. In my experience, every quote has stipulated that the well company would not be responsible for quantity. Once you prove to the customer that it is an industry standard to exclude quantity, you should have made your point. If not, get more estimates. Once you have collected a handful of quotes, all of them excluding water quantity, your customer should give in.

Flow rate

I've had customers demand that I guarantee them a minimum flow rate of recovery for their wells. I've never dealt with a driller who was willing to do this. Therefore, I've never done it. This problem is similar to the issue of quantity. If you can't find a driller who is going to give you a guarantee, you shouldn't offer such a commitment to a customer. Even if I found a driller who would make such a claim, I don't think I would be comfortable doing so.

Quality

Water quality is another issue that most well drillers can't guarantee. The attitude of most well drillers is such that they feel their job is done once water is hit. Most drillers go deep enough to provide a reasonable rate of recovery, but they won't refund your money if the water smells like rotten eggs. Water with a high sulfur content stinks something terrible, and some wells are full of this disgusting water.

Sulfur is not the only disagreeable element found in water. Iron can be a big problem. It stains plumbing fixtures and leaves a black build-up in toilet tanks, water heaters, storage tanks, and so forth. Hard water is common in some areas. Soap does not perform well when mixed with hard water, making it difficult to wash dishes, clothes, and hair. Acid can be so dominant in drinking water that it eats holes in copper pipes and causes upset stomachs in some people. Other minerals, too, can cause homeowners to be disappointed with new wells.

The types of water conditions we are discussing don't usually make water unsafe to drink. But, your customers won't be happy smelling the odor of rotten eggs every time they drink a glass of water. For this reason, you need to cover the bases on water quality in your quotes and contracts. Again, I would use the quotes of well drillers as evidence that a guarantee of quality just isn't standard procedure within the industry.

Location

Location might be an issue that comes to haunt you when installing a well. If the only place to put a well is smack dab in someone's front lawn, you might have a problem. Dug wells have a large diameter and usually consist of a concrete casing and top. It's a lawn ornament that won't get a home on the cover of a fashionable magazine. Drilled wells are not as conspicuous, but they are still not a thing of beauty. The six-inch steel casing of a drilled well is easier to camouflage than the three-foot diameter of a dug well, but you'd better clear the location with your customer before you make a firm commitment.

I suggest that you determine the proposed well location during your site visit. Have the customer agree to the location, then identify some landmarks and take measurements to detail the location on paper. Also, include some option for yourself in case the proposed location proves unsuitable.

Access

Well-drilling rigs are big. It takes a lot of room to move them around. Tree limbs or entire trees might have to be removed to allow access for these big rigs. Make sure you talk this over with your customers in advance. Have the customers agree to whatever is needed to get a well truck in and out.

Permission

Getting permission from a governing body to install a well is normally not a big deal. However, you could run into a problem similar to the one we talked about with septic systems. If a public water supply is available, you might be required to hook up to it. You seldom run into this type of a problem, but don't assume that it couldn't come up.

The trench

Who is going to be responsible for the trench that runs from the well to the house? Well drillers sometimes take care of the trench if they are installing the pump system, but they don't normally provide a trench in the drilling price. If you have full responsibility for all aspects and costs of building a home, you must have the trench dug. That means there is no need to bother the customer with questions about who is expected to do the work. However, if you are acting as the general contractor on only parts of the construction process, the trench could be an expensive issue to have to settle at some point.

The pump system

The pump system falls into a category similar to the trench. If you're taking care of the whole job, the customer doesn't have to know if the pump system is to be installed by a plumber or a well driller. But, if your job is segmented, you might have to determine who is assuming responsibility for the pump system. This work involves a good deal of material and labor, so it is not a cheap place to make a mistake.

A spec sheet

A detailed spec sheet should accompany all of your proposals and contracts. This is just good business for any contractor. Many of the risk elements that we have discussed can be covered in the spec sheet. For example, in the specifications for a well system, you should list all of the types of materials that are to be used. The list would start with the type of well being priced. The type of casing and grouting should also be included. You would then note the brand and size of the pump to be supplied. Reference should be made to the brand, type, and capacity of the pressure tank. The more details you put on your specifications sheet, the better.

As a part of my company policy, I require customers to sign more than just my contracts. I ask them to sign the specifications sheets and any blueprints or drawings that are used. This eliminates any possibility of the customer coming back and claiming that they never saw the documents. There have been occasions when this practice saved me a lot of trouble. If a customer starts complaining about something that is covered in the documents, a quick reminder of their signatures settles them down fast.

What you don't know

What you don't know can definitely hurt you when bidding jobs. If you are not familiar with septic systems and wells, spend some time studying them. You have to be able to talk intelligently about these subjects even though you are not expected to be an expert on the topic. If a customer asks you the difference between a one-pipe jet pump and a two-pipe jet pump, you had better have a good answer. What would you say if a customer wanted your advice on whether to use a jet pump or a submersible pump? Would you recommend using a garbage disposer with a septic tank? You should be prepared to answer such questions with solid answers.

A lot of builders don't know much about wells or septic systems. This is understandable, but not acceptable. You owe it to your customers and yourself to become educated on the services you offer. Failure to do this can hurt your reputation and your bank account. Since you are reading this book, you are obviously interested in learning more about rural water and waste systems. I applaud you for this. Now, let's move onto the next chapter.

6

Soil studies, septic designs, and code-related issues

Soil studies, septic designs, and code-related issues all come into play when builders are working in areas that are not served by public sewers. We've already touched briefly on these issues, but it is now time to dig into the details.

As a builder, it is not your job to do soils studies. These studies are done by soils engineers or county officials. Drawing septic designs isn't part of your job description either. But, you are going to have to know how to interpret them. Code-related issues are the responsibility of the contractor you hire to install your septic systems. However, if you don't have a cursory knowledge of code issues, you might find yourself feeling very foolish. So, what are we going to do about this? Well, I'm going to prepare you with enough background so that you can hold your own with any builder when it comes to talking turkey about septic systems.

Septic designs

I think we should start our discussion with septic designs. Chronologically, soil studies come first, but it might help you to understand the soil studies if you first have knowledge of design criteria. In order to be very accurate here, I'm going to use the actual septic design for my personal home as our example. This design was created by an engineering firm, and the data on it is consistent with design requirements in most areas.

As we run through the information required on a septic design, you are going to see that there are differences in the requirements for a new system and a replacement system. As a builder, you most often are going to be dealing with new systems. However, you might become involved in the utilization of a building lot that once supported a house and septic system. If, for example, the home was destroyed by fire, the new house you are contracted to build might require a replacement septic system. For this reason, I am going to go over both areas of the design form.

The top section of the design form calls for information pertaining to the job location and the property owner's name and address. This information is located at the top left corner of the form. In the right corner is an area for the septic permit to be attached once the design has been approved for a permit.

As we move into the main part of the design, there are many individual boxes that request pertinent information. For example, there is a box where you must indicate if the system is a new system, a replacement system, an expanded system, or an experimental system.

The next question box is one to be filled out by a code official. It asks if the system complies with rules, if it is connected to a sanitary sewer, if the system design is recorded and attached, and if system is installed.

If you are installing a replacement system, you must provide information that states when the failing system was installed. You must also indicate if the failing system is of a bed, trench, or chamber type. If it is another type, you must provide a description.

There is a question that asks for the size of the property. In my case, this was figured in acres, but some lot sizes would be given in actual dimensions. In addition to the lot size, it is necessary to identify the type of zoning in which the property is located. For example, my property is designated as shoreland zoning, since I have frontage on a river.

The next big box of questions deals with what the application requires. The first question asks if a rule variance is required. Is a new system variance needed? Is a replacement variance needed? Subquestions ask if a variance requires approval from a local plumbing inspector or from both a state and local plumbing inspector. The final question in this box asks if a minimum lot size variance is needed?

Moving along, the next box deals with the type of building being served. Is the building a single-family home. You must indicate if the dwelling is of a modular or mobile design. If the building is to be used as a multifamily dwelling, it must reported. Buildings used for purposes other than those already discussed must be described.

As you shift to the right side of the design form, there is a box that asks if the system is engineered, nonengineered, or primitive (referring to the use of an alternative toilet). The next step is to determine the individually installed components of the system, such as a treatment tank, a holding tank, and alternative toilet, a nonengineered disposal area, or a separated laundry system. Just below this question box is another that requires you to describe the type of water supply used for the property.

Question boxes along the bottom of the first page of a design report start with questions about the type of treatment tank to be used. Is the tank aerobic or septic? What is its capacity? Is the tank a standard style, or is it a low-profile unit?

The next box asks about soil conditions, such as the soil profile and condition. Another question asks for the depth limiting factor. In my case, due to bedrock, the depth limiting factor at the septic area was 36 inches.

A third box in the bottom section asks about water conservation. Is there any special procedures used to conserve water? There is a box to check off if a house is equipped with low-volume toilets. There is a place to indicate that separate facilities are used for laundry waste. If alternative toilets are used, they have a box all their own to be checked.

When you enter the box pertaining to the size of a design, you are given five options. They are small, medium, medium-large, large, and extra-large.

The next box deals with septic systems that require pumping. A box must be checked to indicate if a pump is needed or not. Actually, there are three boxes to check, one if a pump is not required, another if a pump might be required, and a third for when a pump is definitely required.

The next to last box on the first page asks about the type and size of the disposal area. Is the system to be made with a bed layout? If so, what is the size of the bed in square feet? Are chambers needed? Are you planning a trench system? If so, how many linear feet are required? For other types of systems, you must specify your plans.

The last box on the first page wants to know the criteria used for design flow. Options include the number of bedrooms, seating capacity, employees, and water records, depending upon the type of building the system is to serve. In the case of a residence, the number of bedrooms is most often used as a guide to determining the number of gallons that are introduced into the system on any given day.

The very bottom of the first page provides a statement regarding an on-site inspection. This section must be dated and completed by the individual who designed the system. Then, you move onto page two.

Page two of a septic design has the top half of the page created with grid boxes. This grid system is supplied so that the design professional can draw a site plan to scale. Part of the site plan shows the location of the building being served, its well (if one is to be installed), and the septic system. Other information might be included in the drawing, such as roads, rivers, ponds, property boundaries, and so forth.

The bottom section of the second page deals with soil descriptions and classifications as they were determined at the observation holes. The first question asks if the hole was a test pit or if it was created by boring. Subjects covered for the soil include texture, consistency, color, and mottling.

The box provided for soil data is ruled to allow for depths ranging from zero to 50 inches. In the case of my tests, the soil texture was sandy loam to a depth of 36 inches, where bedrock was encountered. The consistency was friable throughout. Color ran from brown in the first 6 inches to reddish in the 6- to 16-inch depths, to light brown in the final depths.

Additional information in the bottom section states the soil profile, which in my case was a two. The soil classification was condition "A." My slope was 10 to 15 percent, and my limiting factor was 36 inches, due to bedrock.

Page three of my septic design consists mostly of a detailed drawing of the septic layout. It mandates a low-profile, 1000-gallon septic tank, and no pump. All of the details of the septic system, including the septic tank, the distribution box, and the bed are drawn to scale.

Directly below the drawing of my septic system are some fill-in-the-blank spaces. The first one indicates the depth of fill required on the upslope. In my case, this was 12 inches. For depth of fill on the downslope, I was required to have between 30 and 60 inches.

Construction elevations are also given in this section of the report. My reference elevation was set at zero and marked with a nail and red flag on a tree. The bottom of my disposal area was set at 72 inches below the benchmark (the nail and red flag). For the top of my distribution lines, a calculation was made for 60 inches below the benchmark.

Two cross-sectional drawings were attached to my design. These drawings showed all the details of the installation. For example, the drawing started at the bedrock and showed the original soil surface. It then showed a 12-inch layer of crushed stone. It indicated a layer of hay and 4-inch perforated pipe. This was covered by a 12-inch layer of sandy type fill. Further details showed the rest of the fill needed to accommodate the slope of my system.

Anyone with a reasonable understanding of construction terms could look at my septic design and see exactly what was going to have to happen to make a satisfactory installation. Even if you are a builder who is not familiar with septic systems, reviewing a septic design brings you up to speed quickly.

Design criteria

Design criteria for septic systems can vary from one jurisdiction to the next. You should always consult local authorities to determine the requirements that are in effect within your region. However, I can give you a broad-brush understanding of how the criteria is often set.

Trench and bed systems

Trench and bed systems are two types of drain fields that do not require the use of chambers. Bed systems are the most common of the two. The design criteria for either of these systems is different from that used to layout a chamber system.

The landscape position normally considered suitable for a trench system should not have a slope of more than 25 percent. A slope greater than this can impair the use of equipment needed to install a system. Bed systems are to be limited to a slope of no more than 5 percent. Keep in mind, the numbers I'm giving you are only suggestions. They do not necessarily represent the requirements in your area.

These systems can be installed on land that is level and well-drained and also on the crests of slopes. Convex slopes are considered the best location. Trench and bed systems should not be installed in depressions or at the bases of slopes where suitable surface drainage is not available.

In terms of texture, a sandy or loamy soil is best suited to trench and bed systems. Gravelly and cobbley soils are not as desirable. Clay soil is the least desirable.

When it comes to structure, a strong, granular, blocky, or prismatic structure is best. Platy or unstructured massive soils are the least desirable.

When you are looking at the color of soil with the intent of installing a trench or bed system, look for bright, uniform colors. Such colors indicate a well-drained, well-aerated soil. Ground that gives a dull, gray, or mottled appearance is usually a sign of seasonal saturation. This makes soil unsuitable for a trench or bed system.

Be careful if you find soil that is layered with distinct textural or structural differences. This might indicate that water movement is going to be hindered, and this is not good.

Ideally, there should be between two and four feet of unsaturated soil between the bottom of a drain system and the top of a seasonally high water table or bedrock.

Check with your local authorities to confirm the information I've given you here. You can also ask the authorities to give you acceptable ratings for percolation tests. If you want to do your own perk tests for reference purposes, you simply dig a hole and fill it with water. Then you watch and wait to see how long it takes the soil to absorb the water. You should measure and note the number of inches of water in the hole when you begin the test and the number of minutes it takes for the water to be absorbed. This type of test has much to do with determining the type of drain field that must be installed. I am going to tell you more about perk tests in a moment.

Mound and chamber systems

Mound and chamber systems can have different design criteria than that used for bed and trench systems. I chose bed and trench systems to use as our example since they are the types of systems most often used. Your local code office or county extension office can provide you with more detailed criteria for all types of systems.

Perk tests

To do them properly, perk tests can't be done quickly. The procedures used in the test tend to be consistent among authorities, but you might run into some variations. I know I have. At any rate, I am going to give you a rundown on what I consider to be the right way to conduct a professional perk test. My feelings are based on past research in professional journals and books.

It is common practice to create at least three test holes for any site. You are looking for an average of the results from each hole. By doing this, you rule out the likelihood of having a false reading from just one hole. The holes should be dug or bored at random locations, not too close together, throughout the septic site.

Each test hole should have a diameter of about 6 inches. The depth of the holes should be about the same as the depth of the drain pipes. If you really want to do the test right, scrape the sides of the holes with something sharp and remove the fallen debris from the bottom of the hole. Next, place between one-half and three-quarters of an inch of crushed stone in the bottom of the hole. You are now ready to do the test.

Fill each hole with at least 12 inches of water. This depth of water should be maintained for a minimum of four hours, but overnight is even better. Realistically though, who is going to run out and pour water in a hole on a constant basis throughout the night? Professionals have special fillers and float kits to maintain their water levels. Do they use them? I've never seen it happen.

The reason for keeping the test holes full of water for so long is to saturate the ground. You are attempting to simulate conditions that are going to exist once a drain field is installed. On occasions when you are dealing with sandy soil that contains no clay, you can skip the soaking stage of the test procedure.

You should take your perk test within 15 hours of saturating the soil. Don't wait more than 30 hours to take your tests. Clean out the bottom of the hole if debris has fallen in on the crushed stone, but leave the stone in place. Fill the hole with 6 inches of water. You can use a wooden yardstick as a measuring tool. It is necessary to leave the yardstick in the hole throughout the test, so if you want to test all three holes at the same time, you need three yardsticks.

A notepad and writing instrument are needed to keep records of the water absorption. You should make entries every time the water level drops by $\frac{1}{16}$ of an inch. During this test, you must monitor and record both the time and the loss of water. For example, if you see the first $\frac{1}{16}$ of water drop after four minutes, note the length of time it has taken to have that quantity of water become absorbed by the soil. You can note drops in the water level on a basis of 30-minute increments, if you like.

Testing should continue until two successive water level drops do not vary by more than $\frac{1}{16}$ of an inch and at least three measurements have been made. When dealing with sandy soil, your water might seep out of the hole in less than 30 minutes. Under these conditions, use a standard measurement increment of 10 minutes to record your notes for a period of one hour. The last water level drop is used to calculate a perk rate. I should note that after each measurement of water drop is made, the hole should be replenished with water to a static level of six inches.

What you are after with a perk test is the amount of time it takes for one inch of water to be absorbed. This is done by dividing the time interval used between measurements by the amount of the last water drop. Once you have a rate from each hole, you average the results to come up with your test result. Your figure could be something like a one-inch drop in water every 40 minutes. This data allows you, or someone, to determine what types of septic systems are suitable for a specific location.

Soil types

What soil types are suitable for an absorption-based septic system? A great many types of soils can accommodate a standard septic system. Naturally, some are better than others. Let's take a few moments to discuss briefly what you should look for in terms of soil types.

The best

What is the best type of soil to have when you want to install a normal septic system? There is not necessarily one particular type of soil that is best. However, several types of soil fall into a category of being very desirable. Gravels and gravel-sand mixtures are some of the best soils. Sandy soil is also very good. Soil that is made up of silty gravel or a combination of gravel, sand, and silt can be considered good. Even silty sand and sand-silt combinations rate a good report card. In all of these soil types, it is best to avoid what is known as fines.

Pretty good

Just as there are a number of good soils, there are several soil types that qualify as pretty good for septic tank installations. Gravel that has clay mixed with it is fairly good, and so is a gravel-sand-clay mixture. The same can be said for sand-clay mixtures. Moving down the list of acceptable soil types, you can find inorganic silts, fine sands, silty or clayey fine sands.

Not so good

Inorganic clay, fat clay, and inorganic silt are not so good when it comes to drainage values. This is also true of micaceous or diatomaceous fine sandy or silty soils. These soils can be used in conjunction with absorption-based septic systems, but the systems must be designed to make up for the poor drainage characteristics of the soils.

Just won't do

Some types of soils just won't do when it comes to installing a standard septic system. Of these types of soils, organic silts and clays are included. So are peat and other soils that have a high organic rating.

I've read some pretty sophisticated books on the subject of soils. Believe me when I tell you that an entire book could be written on the subject of soils alone. Operating on the assumption that you are a builder and don't wish to become a soils scientist, I'm going to spare you all the technical information that is available on various types of soils. As a builder, you must rely on someone else to design your septic systems, so I can see no need to go into great detail on the finer points of soil types.

Code-related issues

You and your septic installer must deal with code-related issues. For example, the code does not allow you to begin the installation of a septic system without first obtaining a permit. Don't take a design and install it before obtaining a permit. The code officer might require changes in the proposed design. These changes are much easier to deal with on paper than in the dirt.

Local codes often control the types of materials that can be used in the construction of a septic system. The code officer, or some other official, has the requirements pertaining to the sizing of a system. This sizing procedure might be based on the number of bedrooms in a house or on an estimated daily use of plumbing facilities. In residential homes, the number of bedrooms is often the yardstick by which sizing requirements are made. Commercial and special-use buildings normally have their septic systems sized by using a table of daily water usage.

What else does the local code have to do with your installation? Inspections are a big part of the job. Some communities require multiple inspections. You might have to get the excavated site inspected before any further work is done. It's almost a given that all the installations must be inspected and approved before they are covered up. A third inspection, if there is one, might be done to see if the system was covered and protected properly.

I talked about zoning in Chapter 4. The requirements set forth by zoning regulations must be observed when installing a septic tank. A code officer should inspect for set back, so you have to make sure that any system you install is in compliance with set-back rules and other zoning requirements.

Who is allowed to install a septic system? Does such a system have to be installed by a licensed plumber? I'm not aware of any jurisdiction that requires a licensed plumber to install septic systems. Some areas don't regulate who can install a septic system. Other jurisdictions might require septic installers to be licensed, but the installers don't necessarily have to be plumbers. In fact, I've only known one plumbing company that installed septic systems.

Septic systems are often installed by site contractors. The same people who clear building lots, dig trenches, and grade finished lots routinely install septic systems. Check with your local code office to

see if there are any local requirements that dictate who can install
septic systems. If you have access to a backhoe and dump truck, you
might be able to install the systems yourself. The work is pretty simple, and the profit from installing a septic system is enough to get any
contractor's attention.

When it comes to code questions, it is always best to pursue answers on a local level. Rules that apply in Maine can be meaningless
in South Carolina. Changes in soil types, topography, and water tables can have a lot to do with the installation of septic systems. Talk
to your local code office to get the facts as they pertain to your particular region.

As a home builder, you can take several approaches to the installation of septic systems. You can ignore the mechanics of septic systems and concentrate on subcontracting the work. Or, you can learn
a lot about the various systems and consider installing them yourself
for extra money. Then, there is the middle ground option, where you
know enough to talk intelligently about septic systems with any installer and code officer, but you choose not to take a hands-on approach to the installation of the systems. Which option should you
choose?

I have taken the middle ground. In the areas that I have built
houses, it would have been acceptable for my building company to
install the septic systems. I chose not to do the installations. Why?
Mostly because of the cost of equipment. As a builder, I use a lot of
subcontractors. My emphasis is on volume, and I use my time to sell
houses, rather than build them stick by stick myself. It's more profitable for me to sell a lot of houses and sub the work out, instead of
trying to grab every piece of work just because I have a work crew
available.

Not all builders operate like I do. Many builders in Maine drive
their own nails. Volume building is not a reality in Maine, so local
builders make the most out of every house they sell. In Virginia,
where I built homes in volume, builders did much more work in the
office and in the sales arena than they ever did in the field. Only you
know your personal circumstances.

It's not difficult to understand the job of installing septic systems.
Some of the labor can be rugged, but the technical end of the job is
pretty simple. A good septic design tells you nearly everything you
need to know. With the aid of a code officer or engineer, you can get
any other questions answered on an individual basis. Unless local
code requirements prohibit you from installing septic systems, you
should be able to take on this phase of your building projects for extra money. The choice is yours.

When we get to Chapter 10, we are going to talk about pipe-and-gravel systems. These systems are quite common and easy to install. Whether you want to learn how to install these systems on your own or just want a good understanding of your next septic job, you are going to enjoy Chapter 10. But, to get a break from septic systems, let's turn to the next chapter and discuss drilled wells.

7

Drilled wells

The most dependable individual water sources, that I know, are drilled wells. These wells extend deep into the earth. They reach water sources that other types of wells can't come close to tapping. Since drilled wells take advantage of water that is found deep in the ground, it is very unusual for drilled wells to run dry. This fact accounts for their dependability. However, dependability can be expensive.

Of all the common well types, drilled wells are the most expensive to install. The difference in price between a dug well and a drilled well can be thousands of dollars. But, the money is usually well spent. Dug wells can dry up during hot, summer months. Also, dug wells are more likely to become contaminated. When all the factors are weighed, drilled wells are worth their price.

A house is a big investment. Like a house, a well can be looked upon as an investment. Nobody likes to spend more money than necessary. Builders don't want to needlessly run up the prices of their projects. Home buyers prefer not to pay outrageous prices for the properties they buy. In an effort to keep costs down, some builders look for the cheapest water source they can find. This can be a big mistake. How good is a house without water? Is it really a bargain to install a cheap well only to have it fail six months later?

I don't want to paint a picture that indicates the only type of well worth using is a drilled well. Dug wells can perform very well. Even driven wells can provide a suitable supply of potable water. Some houses get their water from springs and other natural water sources. Your selection of a water supply is going to depend on many factors. Drilled wells are not always practical. However, deep wells with submersible pumps are hard to beat when you want to guarantee a good supply of drinking water.

When I built houses in Virginia, most of the wells were dug wells. In Maine, drilled wells are much more common than dug wells. Geographic location makes a difference in the type of water source used for a house. Chapter 8 is going to give you plenty of information on dug wells. When we get to Chapter 9, you are going to be shown a host of alternative water sources. For now, we are going to concentrate on drilled wells.

Depth

The depth of a drilled well can vary a great deal. In my experience, drilled wells are usually at least 100 feet deep. Some drilled wells extend 500 feet, or more. My personal well is a little over 400 feet deep. Based on my experience as both a builder and plumber, I've found most drilled wells range between 125 to 250 feet deep. When you think about it, that's pretty deep.

It's hard for some people to envision drilling several hundred feet into the earth. I've had many home buyers ask me how I was going to give them such a deep well. Many people have asked me what would happen if the well driller hits rock. Getting through bedrock is not a problem for the right well-drilling rig. While we are on the subject of drilling rigs, let's talk about the different ways to drill a well.

Well-drilling equipment

Various types of well-drilling equipment are available. Each rig type has advantages and disadvantages. As a builder, it can be helpful to know what your drilling options are for different types of sites. There are two basic types of drilling equipment.

Rotary drilling equipment is very common in my area. This type of rig uses a "bit" to bore its way into the earth. The bit is attached to a drill pipe. Extra lengths of pipe can be added as the bit cuts deeper into the ground. The well hole is constantly cleaned out.

Percussion cable tool rigs make up the second type of drilling rig. These drilling machines use a bit that is attached to a wire cable. The cable is repeatedly raised and then dropped to create a hole. A bailer is used to remove debris from the hole.

We could go into a lengthy discussion about all the various types of well rigs available. But, since you are a builder and probably have no desire to become a well driller, there seems to be little point in delving into all the details of drilling a well. What you should know, however, is that there are several types of drilling rigs in existence. Your regional location might affect the types of rigs that are used. A phone call to a few professional drillers can make you aware of your options.

The basics

Let's go over the basics of what is involved, from a builder's point of view, when it comes to drilling a well. Your first step is to decide on a well location. The site of a proposed house, of course, has some bearing on where you want the well to be drilled. Local code requirements address issues pertaining to water wells. For example, the well has to be kept at some minimum distance from a septic field, assuming that a septic system is used.

Who decides where a well should go? Once local code requirements are observed, the decision for a well location can be made by a builder, a home buyer, a well driller, or just about anyone else. If you're building spec houses, the decision is up to you and your driller. Buyers of custom homes might want to take an active role in choosing a suitable well site. As an experienced builder, I recommend that you consult with your customer regarding the well location. Some people are adamant about where the well should be placed.

I often allow my customers to choose the well sites. Many of the home buyers leave the decision up to me. Sometimes a customer chooses a site that is not practical. For example, a well-drilling rig might not be able to easily access the location. When a customer makes what I perceive to be a bad decision, I offer recommendations. Working out an agreeable location is rarely a problem.

I never leave the well location to the discretion of well drillers. Some drillers take the path of least resistance when installing a well. This can result in some very unpleasant well sites. Drilled wells are not as obtrusive as dug wells, but they still don't make good lawn ornaments. If you don't want to arrive on your job site to find a freshly-drilled well in the middle of the front lawn, don't allow well drillers to pick their sites at random.

An experienced driller can provide you with a lot of advice when it comes to picking a spot for a well. Don't overlook this opportunity. It always pays to listen to experienced people. Once you or your customer have chosen a well site, talk the decision over with your driller. You might find that there is a good reason to move from the intended site to a more suitable one. While I feel that you, or a trusted supervisor, should take an active part in picking a well location, I don't think that you should go against the recommendations of a seasoned well driller. If a reputable driller advises you to choose a new location, you should seriously consider following that advice.

Access

Access is one of the biggest concerns a builder has in the well-drilling process. It is the builder's responsibility to provide access for the rigs. Drilling rigs require a lot of room to maneuver. A narrow, private drive with overhanging trees might not be suitable for a well rig.

Having enough width and height to get a well truck into a location is not the only consideration. Well rigs are very heavy. In fact, the ground that these rigs drive over must be solid. New construction usually requires building roads or driveways. If you can arrange to have a well installed along the roadway, your problems with access are reduced.

It is not always possible to install a well alongside a driveway. This might not cause any additional concern, but it could. If the ground where you are working is dry and solid, a well rig can drive over it without any problem. But, a big truck won't be able to go across ground that is wet and muddy, or too sandy. You must consider this possibility when planning a well installation.

Some builders install wells before they build houses. Others wait until the last minute to install a well. Why do they wait? They do it to avoid spending the money for a well before it's needed. This reduces the interest they pay on construction loans and keeps their operating capital as high as possible. When the well is installed last, it can often be paid for out of the closing proceeds from the sale of a house.

I have frequently waited until the end of a job before installing the well. There is some risk to this method. It's possible, I suppose, that a house could be built on land without water. This would truly be a mess. More likely is the risk that the house construction blocks the path of a well rig.

Don't build obstacles for yourself. Like the old joke about painting yourself into a corner, you can build yourself right out of room. When confirming the location of a well, make sure that you can get drilling equipment to the site. I would not enjoy telling the buyers of a custom home that their beautiful shade trees must be cut down to get a well rig to the site. It is safer to install wells before the foundations are built.

Working on your site

Once you have a well driller working on your site, there's not much for you to do but wait. The drilling process can sometimes be completed in a single day. Sometimes, however, the rig must work for longer periods of time. Depending on the type of drilling rig being used, a pile of debris is going to be left behind. Usually, this debris doesn't amount to much and your site contractor can take care of the pile when preparing for finish grading.

Well drillers usually drill a hole suitable for a 6-inch steel casing. The casing is installed to the necessary depth. Once bedrock is penetrated, the rock becomes the well casing. How much casing you need affects the price of the well. Obviously, the less casing that is needed, the lower the price.

Your well driller grouts the well casing as needed to prevent groundwater from running through the casing and dropping down the well. Normally, this procedure is a code requirement. It limits the risk of contamination in the well from surface water. A metal cap is installed on the top edge of the casing, and the well driller's work is done.

Pump installers

Many well drillers offer their services as pump installers. You might prefer to have the driller install the pump, or maybe you would rather have your plumber do the job. A license might be needed in your area for pump installations. Any master plumber can make a pump installation, but you should check to see if drillers are required to have special installation licenses. If they are, make sure that any driller you allow to install a pump is properly licensed. Use a checklist to see that all installation work is going according to plan (Fig. 7-1A and 7-1B).

If you want your driller to install a pump set-up as soon as the drilling is done, you must make arrangements for a trench to run the water service pipe. The cost of this trench is usually not included in prices quoted by pump installers. Watch out for this one, because it is an easy way to lose a few hundred dollars if you don't plan on the cost of the trench.

After your well is installed, you should take some steps to protect it during construction. Heavy equipment could run into the well casing and cause damage. I suggest that you surround the well area with some type of highly visible barrier. Colored warning tape works well, and it can be supported with nothing more than some tree branches stuck in the ground. A lot of builders don't take this safety precaution, but they should. A bulldozer can really do a number on a well, and a well casing can sometimes be difficult for an operator to see.

How far is the pump installer going to take the job? Is the water service pipe going to be run just inside the home's foundation and left for future hook up? Who is going to run the electrical wires and

This check list is intended to help in making reliable submersible pump installations. Other data for specific pumps may be needed.

1. Motor Inspection

____ A. Verify that the model, HP or KW, voltage, phase and hertz on the motor nameplate match the installation requirements. Consider any special corrosion resistance required.

____ B. Check that the motor lead assembly is tight in the motor and that the motor and lead are not damaged.

____ C. Test insulation resistance using a 500 or 1000 volt DC megohmmeter, from each lead wire to the motor frame. Resistance should be at least 20 megohms, motor only, no cable.

____ D. Keep a record of motor model number, HP or KW, voltage, date code and serial number.

2. Pump Inspection

____ A. Check that the pump rating matches the motor, and that it is not damaged.

____ B. Verify that the pump shaft turns freely.

3. Pump/Motor Assembly

____ A. If not yet assembled, check that pump and motor mounting faces are free from dirt and uneven paint thickness.

____ B. Assemble the pump and motor together so their mounting faces are in contact, then tighten assembly bolts or nuts evenly to manufacturer specifications. If it is visible, check that the pump shaft is raised slightly by assembly to the motor, confirming impeller running clearance.

____ C. If accessible, check that the pump shaft turns freely.

____ D. Assemble the pump lead guard over the motor leads. Do not cut or pinch lead wire during assembly or handling of the pump during installation.

4. Power Supply and Controls

____ A. Verify that the power supply voltage, hertz, and KVA capacity match motor requirements.

____ B. Use a matching control box with each single phase three wire motor.

____ C. Check that the electrical installation and controls meet all safety regulations and match the motor requirements, including fuse or circuit breaker size and motor overload protection. Connect all metal plumbing and electrical enclosures to the power supply ground to prevent shock hazard. Comply with National and local codes.

5. Lightning and Surge Protection

____ A. Use properly rated surge (lightning) arrestors on all submersible pump installations unless the installation is operated directly from an individual generator and/or is not exposed to surges. Motors 5HP and smaller which are marked "Equipped with Lightning Arrestors" contain internal arrestors.

____ B. Ground all above ground arrestors with copper wire directly to the motor frame, or to metal drop pipe or casing which reaches below the well pumping level. Connecting to a ground rod does not provide good surge protection.

6. Electrical Cable

____ A. Use cable suitable for use in water, sized to carry the motor current without overheating in water and in air, and complying with local regulations. To maintain adequate voltage at the motor, use lengths no longer than specified in the motor manufacturer's cable charts.

____ B. Include a ground wire to the pump if required by codes or surge protection, connected to the power supply ground. Always ground any pump operated outside a drilled well.

7. Well Conditions

____ A. For adequate cooling, motors must have at least the water flow shown on its nameplate. If well conditions and construction do not assure this much water flow will always come from below the motor, use a flow sleeve as shown in the Application, Installation & Maintenance Manual.

____ B. If water temperature exceeds 30 degrees C (86 °F), reduce the motor loading or increase the flow rate to prevent overheating, as specified in the Application, Installation & Maintenance Manual.

8. Pump/Motor Installation

____ A. Splice motor leads to supply cable using electrical grade solder or compression connectors, and carefully insulate each splice with watertight tape or adhesive-lined shrink tubing, as shown in motor or pump installation data.

____ B. Support the cable to the delivery pipe every 10 feet (3 meters) with straps or tape strong enough to prevent sagging. Use pads between cable and any metal straps.

____ C. A check valve in the delivery pipe is recommended, even though a pump may be reliable without one. More than one check valve may be required, depending on valve rating and pump setting. Install the lowest check valve below the lowest pumping level of the well, to avoid hydraulic shocks which may damage pipes, valve or motor.

____ D. Assemble all pipe joints as tightly as practical, to prevent unscrewing from motor torque. Recommended torque is at least 10 pound feet per HP (2 meter-KG per KW).

____ E. Set the pump far enough below the lowest pumping level to assure the pump inlet will always have at least the Net Positive Suction Head (NPSH) specified by the pump manufacturer, but at least 10 feet (3 meters) from the bottom of the well to allow for sediment build up.

7-1A *Submersible pump installation checklist.* A.Y. McDonald MFG. Co., Dubuque, Iowa

make the electrical connections? Does the installation price include a pressure tank and all the accessories needed to trim it out? Who installs the pressure tank? You need answers to these questions before you award a job to a subcontractor. Get detailed information from the pump installer on the pump being installed (Figs. 7-2A and 7-2B). Pump systems involve a lot of steps and materials. It's easy for a contractor to come in with a low bid by shaving off some of the work re-

____ F. Check insulation resistance from dry motor cable
ends to ground as the pump is installed, using a 500
or 1000 volt DC megohmmeter. Resistance may
drop gradually as more cable enters the water, but
any sudden drop indicates possible cable, splice or
motor lead damage. Resistance should meet motor
manufacturer data.

9. After Installation

____ A. Check all electrical and water line connections and
  ·   parts before starting the pump. Make sure water
delivery will not wet any electrical parts, and
recheck that overload protection in three phase
controls meets requirements.

____ B. Start the pump and check motor amps and pump
delivery. If normal, continue to run the pump until
delivery is clear. If three phase pump delivery is
low, it may be running backward because phase
sequence is reversed. Rotation may be reversed
(with power off) by interchanging any two motor
lead connections to the power supply.

____ C. Connect three phase motors for current balance
within 5% of average, using motor manufacturer
instructions. Unbalance over 5% will cause higher
motor temperatures and may cause overload trip,
vibration, and reduced life.

____ D. Make sure that starting, running and stopping cause
no significant vibration or hydraulic shocks.

____ E. After at least 15 minutes running, verify that pump
output, electrical input, pumping level, and other
characteristics are stable and as specified.

Date _____   Filled In By

10. Installation Data

Well Identification _____

Check By _____

Date _____ / _____ / _____

Notes _____

_____

_____

_____

_____

_____

_____

_____

_____

7-1B *Submersible pump installation checklist.* A.Y. McDonald MFG. Co., Dubuque, Iowa

sponsibilities in the fine print. Be careful, or you might wind up paying a lot more than you planned for a well system.

We have now covered, in an abbreviated form, the installation of
a drilled well and a pump system. However, you should be aware of
some particulars so that you can better supervise the work. Let's concentrate on these issues now.

RMA No. _____

INSTALLER'S NAME _____   OWNER'S NAME _____

ADDRESS_____   ADDRESS _____

CITY _____ STATE_____ ZIP_____   CITY _____ STATE_____ ZIP_____

PHONE (____) _____ FAX (____) _____   PHONE (____) _____FAX (____) _____

CONTACT NAME _____   CONTACT NAME _____

WELL NAME/ID_____ DATE INSTALLED_____

MOTOR:

Motor No. _____ Date Code _____ HP _____ Voltage _____ Phase _____

PUMP:

Manufacturer _____ Model No. _____ Curve No. _____ Rating: _____ GPM@_____ft. TDH

NPSH Required: _____ ft. NPSH Available:_____ ft. Actural Pump Delivery_____GPM@ _____ PSI

Operating Cycle: _____ON (Min./Hr.) _____ OFF (Min./Hr.) (Circle Min. or Hr. as appropriate)

YOUR NAME _____ DATE ____/_____/_____

WELL DATA:

Total Dynamic Head _____ft.
Casing Diameter_____in.
Drop Pipe Diameter_____in.
Static Water Level _____ft.
Drawdown (pumping) Water Level_____ft.

Checkvalves at _____&_____&
_____&_____ft.
❏ Solid ❏ Drilled

Pump Inlet Setting _____ ft.
Flow Sleeve: ____No____ Yes, Dia._____in.

Casing Depth_____ft.
❏ Well Screen ❏ Perforated Casing
From_____to_____ft. & _____to_____ft.

Well Depth_____ft.

TOP PLUMBING:
Please sketch the plumbing after the well head (check valves, throttling valves, pressure tank, etc.) and indicate the setting of each device.

Form No. 2207 2/94

7-2A *Submersible motor installation record.* A.Y. McDonald MFG. Co., Dubuque, Iowa

Quantity and quality

Few (if any) well drillers make commitments when it comes to the quantity and quality of water produced by a well. Every well driller I've talked to has refused to guarantee the quantity or quality of water. The only guarantee that I've been able to solicit has been an as-

POWER SUPPLY:

Cable: Service Entrance to Control _____ft. _____ AWG/MCM ❑ Copper or ❑ Aluminum, ❑ Jacketed or ❑ Individual Conductors

Cable: Control to Motor _____ft. _____ AWG/MCM ❑ Copper or ❑ Aluminum, ❑ Jacketed or ❑ Individual Conductors

Transformers:

KVA _____ #1 _____ #2 _____ #3

Intial Megs (motor & lead)   T1 _____ T2 _____ T3 _____
Final Megs (motor, lead & cable)   T1 _____ T2 _____ T3 _____

Incoming Voltage:

No Load: L1-L2 _____ L2-L3 _____ L1-L3 _____
Full Load: L1-L2 _____ L2-L3 _____ L1-L3 _____

Running Amps:

HOOKUP 1:
Full Load: L1 _____ L2 _____ L3 _____
        %Unbalance _____
HOOKUP 2:
Full Load: L1 _____ L2 _____ L3 _____
        %Unbalance _____
HOOKUP 3:
Full Load: L1 _____ L2 _____ L3 _____
        %Unbalance _____

Control Panel:

Panel Manufacturer: _____

Short Circuit Device: ❑ Circuit Breaker: Rating _____ Setting _____
            ❑ Fuses: Rating _____ Type _____
            ❑ Standard ❑ Delay

Starter Manufacturer: _____ Starter Size _____

Type of Starter: ❑ Full Voltage   ❑ Autotransformer
         ❑ Other: _____ Full Voltage in _____ sec.

Heaters Manufacturer: _____

Number _____ Adjustable Set at: _____ amps.

Subtrol-Plus: ❑ No ❑ Yes: Registration No. _____

If yes, Overload Set?   ❑ No ❑ Yes Set at _____ amps.
    Underload Set?   ❑ No ❑ Yes Set at _____ amps.

Controls are Grounded to: ❑ Well Head ❑ Motor
            ❑ Rod   ❑ Power Supply

Ground Wire Size: _____ AWG/MCM

Comments: _____

_____
_____
_____
_____
_____
_____
_____
_____
_____
_____

7-2B *Submersible motor installation record.* A.Y. McDonald MFG. Co., Dubuque, Iowa

surance that the driller would hit water. Builders have to be aware of this issue.

A customer might ask you, the builder, to specify what the flow rate of their new well is going to be. It is beyond the ability of a builder to make such a prediction. An average, acceptable well could have a 3-gallon-per-minute (gpm) recovery rate. Another well might replenish itself at a rate of 5 gpm. Some wells have much faster recovery rates, others with slower rates. Anything below 3 gpm is less than desirable, but it can be made to suffice.

How can you deal with the recovery rate? You really can't. Sometimes, by going deeper, a well driller can hit a better aquifer that produces a higher rate of recovery, but there is no guarantee. So, don't make a guarantee to your customers. Show your customers the disclaimers on the quotes from well drillers and use that evidence to back up your point that there is no guarantee of recovery rate in the well business. Now, I could be wrong. You might find a driller who guarantees a rate, but I've yet to meet one.

Even though you can't know what a recovery rate is going to be when you start to drill a well, you can determine what it is after water has been hit. Your well driller should be willing to test and establish your recovery rate. Every driller I've ever used has performed this service. You need to know what the recovery rate is in order to size a pump properly.

It is also important that you know the depth of the well, and this is something that your driller can certainly tell you. As you gather information, record it for future reference. The depth of the well is also a factor when selecting and installing a pump. Don't let your driller leave the job until you know the well depth and the recovery rate. Guidelines for minimum water requirements (Figs. 7-3 and 7-4) are available in your local plumbing code and from pump manufacturers.

If you're the type of person who likes to do a little engineering on your own, you can refer to information provided by pump manufacturers to calculate your pump needs (Figs. 7-5 and 7-6).

Quality

The quality of water is difficult to determine when a well is first drilled. It can take days, or even weeks, for the water in a new well to assume its posture. In other words, the water you test today could offer very different results when tested two weeks from now. Before a true test of water quality can be conducted, it is often necessary to disinfect a new well. Many local codes require disinfection before testing for quality.

Wells are usually disinfected with chlorine bleach. Local requirements on disinfection vary, so I won't attempt to tell you an exact procedure. In general, a prescribed amount of bleach is poured into a well. It is allowed to sit for some specific amount of time, as regulated by local authorities. Then the well pump is run to deplete the water supply in the well. As a rule of thumb, the pump is run until there is no trace of chlorine odor in the tap water. When the well re-

Average water requirements for general
service around the home and farm

Each person per day, for all purposes	75 gal.
Each horse, dry cow, or beef animal	12 gal.
Each milking cow	35 gal.
Each hog per day	4 gal.
Each sheep per day	2 gal.
Each 100 chickens per day	4 gal.

Average amount of water required by
various home and yard fixtures

Drinking fountain, continuously flowing	50 to 100 gal. per day
Each shower bath	Up to 30 gal. @ 3–5 gpm
To fill bathtub	30 gal.
To flush toilet	6 gal.
To fill lavatory	2 gal.
To sprinkle 1/4" of water on each 1000 square feet of lawn	160 gal.
Dishwashing machine — per load	7 gal. @ 4 gpm
Automatic washer —per load	Up to 50 gal. @ 4–6 gpm
Regeneration of domestic water softener	50–100 gal.

Average flow rate requirements by
various fixtures
(gpm = gal. per minute; gph = gal. per hour)

Shower	3–5 gpm
Bathtub	3–5 gpm
Toilet	3 gpm
Lavatory	3 gpm
Kitchen sink	2–3 gpm
1/2" hose and nozzle	200 gph
3/4" hose and nozzle	300 gph
Lawn sprinkler	120 gph

7-3 *Average water requirements for general service.*
A.Y. McDonald MFG. Co., Dubuque, Iowa

plenishes itself, the new water in the well should be ready for testing. But, again, check with your local authorities for the correct procedure to use in your area. By the way, the builders are usually responsible for conducting the disinfection process.

ENGINEERING DATA
Formula: GPD ÷ 2 = Capacity

1. The first important step in planning for a sufficient and economical water supply is to estimate the actual capacity required. It is a common mistake to underestimate the real need. Dissatisfaction and higher cost usually occur as a result of doing the job over. This can be eliminated by proper planning.

   A general average for each member of a household, for all purposes, including kitchen, laundry, bath, and toilet (but not including yard fixtures) is 75 gallons per day per person. It is suggested the total requirement of a 24-hour day be pumped in two (2) hours for two basic reasons:

   a. To take adequate care of the peak period demand such as several outlets on at one time, and

   b. To provide an economical pump selection.

   Added to the normal daily household requirement should be the demand required for any special requirement such as sprinkling systems, livestock, etc. Usually manufacturers will rate the water demand for specific special items such as milk can washers separately. The normal livestock requirements are as follows:

	Approx. Gallons Per Day
Each horse	12
Each producing cow	15
Each nonproducing cow	12
Each producing cow with drinking cups	30–40
Each nonproducing cow with drinking cups	20
Each hog	4
Each sheep	2
Each 100 chickens	4–10
Yard fixtures:	
½-inch hose with nozzle	200
¾-inch hose with nozzle	300
Bath houses	10
Camp	
Construction, semipermanent	50
Day (with no meals served)	15
Luxury	100–150
Resorts (day and night, with limited plumbing)	50
Tourists with central bath and toilet facilities	35
Cottages with seasonal occupancy	50
Courts, tourists with individual bath units	50
Clubs	
Country (per resident member)	100
Country (per nonresident member present)	25

7-4 *Daily water requirements.* A.Y. McDonald MFG. Co., Dubuque, Iowa

Dwellings

Luxury	75
Multiple family, apartments (per resident)	60
Rooming houses (per resident)	50
Single family	75
Estates	100–150
Factories (gallons/person/shift)	15–35
Institutions other than hospitals	75–125
Hospitals (per bed)	250–400
Laundries, self-serviced (gallons per washing, i.e., per customer)	50

Motels

With bath and toilet (per bed space)	100

Parks

Overnight with flush toilets	25
Trailers with individual bath units	50

Picnic

With bath houses, showers, and flush toilets	20
With only toilet facilities (gal./picnicker)	10
Restaurants with toilet facilities (per patron)	10
Without toilet facilities (per patron)	3
With bars and cocktail lounge (additional quantity)	2

Schools

Boarding	50–70
Day with cafeteria, gymnasiums and showers	25
Day with cafeteria but no gymnasiums or showers	20
Service stations (per vehicle)	10
Stores (per toilet room)	400
Swimming pools	10

Theaters

Drive-in (per car space)	5
Movie (per auditorium seat)	5

Workers

Construction (semipermanent)	50
Day (school or offices per shift)	15

Providing an adequate water supply provides for a healthy family and higher production from livestock. Assuming the total daily requirement is calculated to be 1200 gpd (gallons per day), a pump would be selected for a capacity of 10 gpm (gallons per minute) based on the following formula:

$$1200 \text{ gph} \div 2 \text{ equals } 600 \text{ gph (gal. per hr.)}$$

Example: 5 in family @ 75 gpd each person	375
1¾" hose with nozzle @ 300	300
10 non-producing cows with cups @ 20	200
Total 24 hr. req.	875

$$875 \div 2 = 438 \text{ gph or } 7.3 \text{ gpm pump selection}$$

7-4 *Continued.*

Engineering Data
Drop Cable Selection Chart
Single-phase, two or three-wire cable, 60 Hz (service entrance to motor)

Motor Rating		Copper Wire Size									
Volts	HP	14	12	10	8	6	4	3	2	1	0
115	⅓	130	210	340	540	840	1300	1610	1960	2390	2910
	½	100	160	250	390	620	960	1190	1460	1780	2160
230	⅓	550	880	1390	2190	3400	5250	6520	7960	9690	11770
	½	400	650	1020	1610	2510	3880	4810	5880	7170	8720
	¾	300	480	760	1200	1870	2890	3580	4370	5330	6470
	1	250	400	630	990	1540	2380	2960	3610	4410	5360
	1.5	190	310	480	770	1200	1870	2320	2850	3500	4280
	2	150	250	390	620	970	1530	1910	2360	2930	3620
	3	120	190	300	470	750	1190	1490	1850	2320	2890
	5	0	110*	180	280	450	710	890	1110	1390	1740
	7.5	0	0	120*	200	310	490	610	750	930	1140
	10	0	0	0	160*	250	390	490	600	750	930
	15	0	0	0	0	170*	270	340	430	530	660

1 foot = .3048 meter

Three-phase, three-wire cable, 60 Hz 200 and 230 volts (service entrance to motor)

Motor Rating		Copper Wire Size (1)											
Volts	HP	14	12	10	8	6	4	3	2	1	0	00	000	0000
200V 60 Hz Three-Phase Three-Wire	½	710	1140	1800	2840	4420								
	¾	510	810	1280	2030	3160								
	1	430	690	1080	1710	2670	4140	5140						
	1.5	310	500	790	1260	1960	3050	3780						
	2	240	390	610	970	1520	2360	2940	3610	4430	5420			
	3	180	290	470	740	1160	1810	2250	2760	3390	4130			
	5	110*	170	280	440	690	1000	1350	1660	2040	2490	3050	3670	4440
	7.5	0	0	200	310	490	770	960	1180	1450	1770	2170	2600	3150
	10	0	0	150*	230	370	570	720	880	1090	1330	1640	1970	2390
	15	0	0	0	160	250	390	490	600	740	910	1110	1340	1630
	20	0	0	0	0	190*	300	380	460	570	700	860	1050	1270
	25	0	0	0	0	0	240*	300	370	460	570	700	840	1030
	30	0	0	0	0	0	200*	250*	310	380	470	580	700	850
230V 60 Hz Three-Phase Three-Wire	½	810	1300	2040	3210	4990								
	¾	590	940	1480	2330	3620								
	1	490	790	1240	1960	3050	4720	5860						
	1.5	360	580	920	1450	2260	3510	4360						
	2	280	450	700	1110	1740	2710	3370	4130	5070	6200			
	3	210	340	540	860	1340	2080	2580	3170	3880	4730			
	5	130*	200	320	510	800	1240	1550	1900	2330	2850	3490	4200	5080
	7.5	0	140*	230	360	570	890	1100	1350	1660	2030	2480	2980	3600
	10	0	0	170*	270	420	660	820	1010	1240	1520	1870	2260	2740
	15	0	0	0	180*	290	450	560	690	850	1040	1280	1540	1860
	20	0	0	0	140*	220*	350	430	530	660	810	990	1200	1450
	25	0	0	0	0	180*	280	350	430	530	650	800	970	1170
	30	0	0	0	0	0	230*	290	350	440	540	660	800	970
460V 60 Hz Three-Phase Three-Wire	½	3770	6020	9460										
	¾	2730	4350	6850										
	1	2300	3670	5770	9070									
	1.5	1700	2710	4270	6730									
	2	1300	2070	3270	5150	8050								
	3	1000	1600	2520	3970	6200								
	5	590	950	1500	2360	3700	5750							
	7.5	420	680	1070	1690	2640	4100	5100	6260	7680				
	10	310	500	790	1250	1960	3050	3800	4680	5750	7050			
	15	0	340*	540	850	1340	2090	2600	3200	3930	4810	5900	7110	
	20	0	0	0	650	1030	1610	2000	2470	3040	3730	4580	5530	
	25	0	0	330*	530	830	1300	1620	1990	2450	3010	3700	4470	5430
	30	0	0	270*	430	680	1070	1330	1640	2030	2490	3060	3700	4500
	40	0	0	0	320*	500*	790	980	1210	1490	1830	2250	2710	3290
	50	0	0	0	0	410*	640	800	980	1210	1480	1810	2190	2650
	60	0	0	0	0	0	540*	670	830	1020	1250	1540	1850	2240
	75	0	0	0	0	0	440*	550*	680*	840	1030	1260	1520	1850
	100	0	0	0	0	0	0	0	0	500*	620*	760*	940	1380
	125	0	0	0	0	0	0	0	0	0	600*	740*	890*	1000
	150	0	0	0	0	0	0	0	0	0	0	630*	760*	920*
	175	0	0	0	0	0	0	0	0	0	0	0	670*	810*
	200	0	0	0	0	0	0	0	0	0	0	0	590*	710*

Lengths marked * meet the U.S. National Electrical Code ampacity only for **individual** conductor 75°C cable. Only the lengths **without** * meet the code for **jacketed** 75°C cable. Local code requirements may vary.

CAUTION!! Use of wire sizes smaller than determined above **will void warranty**, since low starting voltage and early failure of the unit will result. Larger wire sizes (smaller numbers) may always be used to improve economy of operation.

(1) If aluminum conductor is used, multiply above lengths by 0.61. Maximum allowable length of aluminum wire is considerably shorter than copper wire of same size.

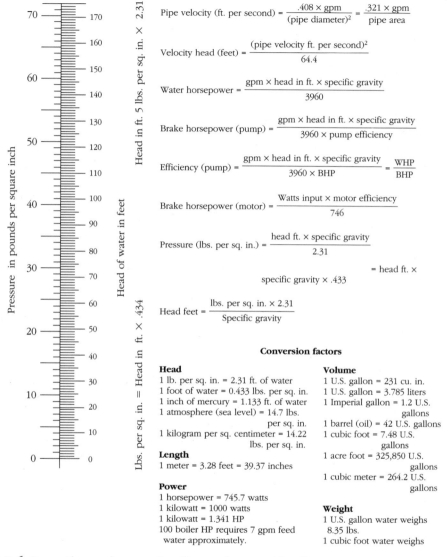

Engineering data
Formulas and conversion factors
Centrifugal pumps

Pipe velocity (ft. per second) = $\dfrac{.408 \times \text{gpm}}{(\text{pipe diameter})^2} = \dfrac{.321 \times \text{gpm}}{\text{pipe area}}$

Velocity head (feet) = $\dfrac{(\text{pipe velocity ft. per second})^2}{64.4}$

Water horsepower = $\dfrac{\text{gpm} \times \text{head in ft.} \times \text{specific gravity}}{3960}$

Brake horsepower (pump) = $\dfrac{\text{gpm} \times \text{head in ft.} \times \text{specific gravity}}{3960 \times \text{pump efficiency}}$

Efficiency (pump) = $\dfrac{\text{gpm} \times \text{head in ft.} \times \text{specific gravity}}{3960 \times \text{BHP}} = \dfrac{\text{WHP}}{\text{BHP}}$

Brake horsepower (motor) = $\dfrac{\text{Watts input} \times \text{motor efficiency}}{746}$

Pressure (lbs. per sq. in.) = $\dfrac{\text{head ft.} \times \text{specific gravity}}{2.31}$
= head ft. × specific gravity × .433

Head feet = $\dfrac{\text{lbs. per sq. in.} \times 2.31}{\text{Specific gravity}}$

Conversion factors

Head
1 lb. per sq. in. = 2.31 ft. of water
1 foot of water = 0.433 lbs. per sq. in.
1 inch of mercury = 1.133 ft. of water
1 atmosphere (sea level) = 14.7 lbs.
per sq. in.
1 kilogram per sq. centimeter = 14.22
lbs. per sq. in.

Length
1 meter = 3.28 feet = 39.37 inches

Power
1 horsepower = 745.7 watts
1 kilowatt = 1000 watts
1 kilowatt = 1.341 HP
100 boiler HP requires 7 gpm feed
water approximately.

Volume
1 U.S. gallon = 231 cu. in.
1 U.S. gallon = 3.785 liters
1 Imperial gallon = 1.2 U.S.
gallons
1 barrel (oil) = 42 U.S. gallons
1 cubic foot = 7.48 U.S.
gallons
1 acre foot = 325,850 U.S.
gallons
1 cubic meter = 264.2 U.S.
gallons

Weight
1 U.S. gallon water weighs
8.35 lbs.
1 cubic foot water weighs

7-6 *Formulas and conversion factors for centrifugal pumps.* A.Y. McDonald MFG.
Co., Dubuque, Iowa

Once the well is ready to test, a water sample is taken from a
faucet in the house. Test bottles and collection instructions are available from independent laboratories. Follow the instructions provided

by the testing facility. As a rule of thumb, remove any aerator that might be installed on the spout of a faucet before taking your water collection. To kill any bacteria that might cling to the faucet and wash into the collection bottle, it is often recommended that you first hold a flame to the spout. Don't attempt this step when testing from a plastic faucet! Many plumbers take water tests from outside hose bibs, and this is fine. A torch can be used to sterilize the end of the hose bib.

Before collecting any water for a test, allow the water to run through the faucet for several minutes. It is best to drain the contents of a pressure tank and have it refilled with fresh well water for the test. You can quickly drain the reserve water by opening the cold water faucet for the bathtub.

Once a water sample is collected, it is taken or mailed to a lab. Time is of the essence when testing for bacteria, so don't let the water bottle ride around in your truck for a few days before it's mailed. Again, follow the instructions provided by the lab for delivering the water.

You can request the lab to perform different types of tests. A mandatory test reveals if the water is safe to drink. If you want to know more, you have to ask for additional tests. For example, you might want to have a test done to see if radon is present. Many wells have mineral contents in sufficient quantity to affect the water quality. Acid levels in the well might be too high. The water could be hard, which causes plumbing fixtures to stain and makes it difficult to wash with soaps. We are going to talk about the effects of various mineral contents in Chapter 16.

Drinking water, or potable water as it can be called, is your primary goal when drilling a well. It is rare to find that water from a drilled well does not test well enough to meet minimum requirements for safe drinking. In fact, I've never known of a drilled well that didn't produce water suitable for drinking. Nevertheless, an official statement of quality acceptability is generally required by code officials and those who loan money for houses.

If you get into a discussion on water quality, you might have to pinpoint exactly what you are talking about. Is the subject only related to the water being safe to drink? Or does it extend into mineral contents and such? Most professionals look at water quality on an overall basis, which includes mineral contents.

Supervising the drilling

You should not have to supervise the drilling of a new well. Standing around all day, watching a well rig drill, is pretty boring and not very productive. Once you have chosen a well location and have made sure the driller is drilling in the proposed area, you can pretty much ignore the drilling process until it's done. Near the end of the job, you might want to make an appearance to find out the height of the well casing and to obtain information of the well depth and flow rate.

Trenching

A trench must be dug for the installation of water service and for the electrical wires that run out to a submersible pump. It is possible to pump water from a deep well with a two-pipe jet pump, but submersible pumps are, in my opinion, far superior. When a submersible pump is used, it hangs in the well water. Electrical wires must be run to the well casing and down into the well. The wires and the water service pipe can share the same trench.

As with any digging, you must make sure that there are no underground utilities in the path of your excavation. Most communities offer some type of underground utility identification service. In many places, only one phone is needed to get someone to mark all the underground utility locations on your work site. However, in some parts of the country, it might be necessary to call individual utility companies. I would expect that, as a builder, you are familiar with the process in your area.

Once you have a clear path to dig, a trench must be dug to a depth that is below the local frost line. The water service pipe has water in it at all times, so it must be buried deep enough to avoid freezing. How deep is deep enough? It varies from place to place. In Maine, I have to get down to a depth of four feet to be safe from freezing. In Virginia, the frost line was set at eighteen inches. Your local code office can tell you what the prescribed depth is in your area.

After a trench is opened up, you can arrange for the pump installer to come to the site. Most installers want to do all of their work in one trip. This means that you must have enough of the house built to allow an installer to bring the water pipe through the foundation and to set the pressure tank. As a helpful hint, you should install a sleeve in your foundation as it is being poured so that the pump installer doesn't have to cut a hole through the foundation.

Check with your local plumbing inspector to determine what size sleeve is needed. Most plumbing codes require a sleeve in a foundation wall to be at least two pipe sizes larger than the water pipe. For a typical 1-inch-diameter well pipe, the sleeve must be at least 2 inches in diameter. But again, check with your local code office, because plumbing codes do vary from place to place.

At this point, you are ready to have your pump system installed. The pump might be installed by your well driller or your plumber. Wiring for the pump should be done by a licensed electrician. All of your subcontractors should, of course, carry liability insurance that protects you and the property owner from their mistakes and accidents.

Installing a pump system

Installing a pump system for a drilled well is not a difficult procedure. It is, however, a job that usually must be done by a licensed professional. Even if you know how to do all of the work, you probably can't do it legally without a license. Since almost everyone uses submersible pumps when dealing with deep wells, I'll base my examples on the assumption that you, too, are using a submersible pump.

Since you are a builder, rather than a pump installer, I won't go into every little detail of a pump installation. I cover all the key points, but without the step-by-step instructions that I would give someone wishing to learn the trade. In other words, I'm not going to waste your time with instructions on applying pipe dope or on how tight to turn a fitting, or how to carry out every other little plumbing process. What I am going to tell you is how to make sure that the installer you choose does a good job.

At the well casing

Let's start our work at the well casing (Fig. 7-7). You should be looking at an empty trench and the side of a steel well casing when a pump installation begins. A hole has to be cut through the side of the well casing to allow a pitless adapter to be installed. A cutting torch can be used to make this hole, but most installers use a metal-cutting hole saw and a drill. The hole size is determined by the pitless adapter. It's important to keep the hole at proper tolerances. If the pitless adapter doesn't fit well, groundwater might leak into the well by getting past the gasket provided with the adapter.

The hole in the side of the well casing should be positioned so that a water pipe laying in the trench is going to line up with the pitless adapter once it is installed. Once the hole has been cut, the pitless adapter needs to be installed. This is really a two-person job. One person works down in the trench, while another person works from above the main opening in the well casing.

A long, threaded pipe is needed to position the piece of the pitless adapter that is installed inside the well. Most plumbers make up

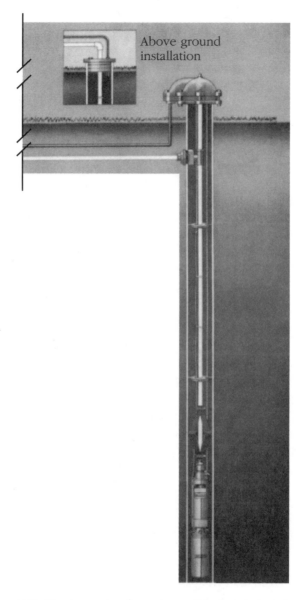

7-7 *Components of a submersible pump installation.* Goulds Pumps, Inc.

a T-shaped pipe tool to use for this part of the job. The pitless adapter
is screwed onto the threads of the T-shaped tool. With this done, one
person lowers the pitless adapter down the well and into position. A
threaded protrusion on the adapter is intended to poke through the

hole in the side of the well casing. When it does, the person in the trench installs a gasket and the other part of the pitless adapter, which is basically a retaining ring. The ring is tightened to create a water-tight seal and to hold the pitless adapter in place. At this point, the T-shaped tool is unscrewed from the pitless adapter and laid aside. As a builder, you should check to see that the pitless adapter is tight and in position to prevent contaminated groundwater from entering around the hole.

Most installers put their pump rigs together next. This involves one of two things—either a truck equipped with a reel system for well pipe or enough room to lay all of the well pipe out in a fairly straight line. Laying out enough pipe can be a problem with deep wells. It is important, however, that the well pipe be laid out and not allowed to kink, unless it is being fed into the well from a reel system.

The pipe used for most wells comes in long coils. It is best to avoid joints in well pipe whenever possible. With the long lengths of pipe available, there should be no reason why a joint would be needed to splice two pieces of pipe together. When joints and connections are made with standard well pipe, it is best to use metallic fittings. Nylon fittings are available, but they might break under stress more quickly than metal fittings.

Polyethylene (PE) pipe is the type used most often for well installations. This black plastic pipe is sold in coils and can be used for both the vertical drop pipe in the well and the horizontal water service in the trench. In houses where hot water is available, PE pipe must not extend more than five feet inside the foundation wall. The plumbing code requires the same pipe type to be used for water distribution of hot and cold water. PE pipe is not rated to handle hot water, so it cannot be used as an interior water distribution pipe.

Other types of pipe can be used for well systems, but PE pipe remains the most popular. This pipe has some drawbacks for installers. The material kinks easily. If the pipe kinks, it should not be used because the bend weakens the pipe. If you see an installer kink a pipe, make sure that the kinked section isn't used. Since couplings are undesirable both in the well and in the trench, a kinked pipe might mean getting a whole new roll of pipe, depending upon where the kink occurs.

Another fault of PE pipe is its tendency to become very hard to work with in cold weather. Fittings, which are an insert type, are difficult to push into the pipe when it's cold. Warming the ends of the pipe with a torch or heat gun makes the material pliable and easier to work. However, care must be used to avoid melting the pipe. All connections made with PE pipe should be made with two stainless-steel hose clamps. One clamp is all that is required by code, but a second clamp provides cheap insurance against leaks.

When a coil of PE pipe is unrolled, it takes some work to straighten it out. This is basically a two-person job, although I have done it alone. The pipe should be stretched out as straight as possible, without kinking, and then it must be manipulated in a looping motion to make it lay flat. This step of the installation process should not be ignored.

Once the well pipe is laying flat, an installer can proceed to assemble the pump and accessories to the pipe. A torque arrestor should be installed on the pipe. This limits vibration in the well as the pump runs. A male, insert adapter (brass please) should be used to connect the pipe to the pump. Another insert adapter connects the pipe to the pitless adapter fitting that has yet to be installed.

Electrical wires must run from the pump to the top of the well casing. Some plumbers use electrical tape to secure electrical wire to the well pipe. This is a common procedure. Other installers use plastic guides that slide over the well pipe to secure the wires. Either way, someone has to make sure that the wires are secure. It is also very important to make sure that the wires are not damaged as the pump assembly is lowered into the well. Waterproof splice kits can be used to joint electrical wires that extend into a well, and of course, waterproof wire is required.

All of the piping and wiring is connected to a submersible pump before the pump is installed. This work is normally done near the well, on the ground. Once the entire assembly is put together, it is lowered into the well. Some companies have special trucks for this part of the job, but a lot of installers do it the old-fashioned way—by hand.

When the pump assembly is put down the well, the work goes much better if two people are involved. Lowering a pump assembly by hand is not difficult. The T-bar tool is once again connected to the wedge-shaped part of a pitless adapter and attached to the upper end of the drop pipe. Smart installers attach nylon rope to the submersible pump as a safety rope. I've seen a number of installers skip this step, but I won't allow it on my jobs. I insist on having a safety rope installed.

Submersible pumps (Figs. 7-8 and 7-9) are held in the well only by the well pipe, unless a safety rope is used. If a fitting pulls loose, an expensive pump can be forever lost in the well. The rope, which is tied to the top of the well casing, gives you a means of retrieving the pump should anything go wrong with the piping arrangement. For the low cost of nylon rope, it's senseless not to put it to use.

Bronze discharge casting

•**External check valve**

Rubber bearing
Maximum abrasion resistance

•**Completely field serviceable**
Replaceable stack assembly

Stainless steel hex shaft
Achieves maximum corrosion resistance for longer life expectancy

Acetal impellers
Tough, lightweight, precision molded

Lexan® diffusers
Abrasion and corrosion resistant

Polished stainless steel shell

•**Noncorrosive screen**

•**Noncorrosive cable guard**

•**Bronze center casting**

•**Stainless steel motor**
2 or 3 wire models, with built-in lightning protection

7-8 *Cutaway of a submersible pump.* A.Y. McDonald MFG. Co., Dubuque, Iowa

The pump is lowered into the well. During this process, the person working the assembly down into the casing must take care not to scrape the electrical wires along the edge or sides of the casing. If the insulation on the wires should be cut by the casing, the pump might fail to operate. Once the pump is in position, the pitless adapter piece is put into place with its mating part that is secured to the well casing. This is done simply by lining the wedge up with the groove in the stationary piece and tapping it into place. Then the safety rope is adjusted and tied to the casing. The T-bar is unscrewed and removed. The only step remaining at the well head is the connection of electrical wires and the replacement of the well cap. An experienced crew can put together and install a pump assembly in about two hours.

The water service

The next step in the installation of a pump system is the water service. This pipe might be the same type used in the well, or it might

■ Capacities to 82 gpm
■ Heads to 590 feet (180m)

■ 1 through 7¹/₂ horsepower models
■ Discharge head and bracket
corrosion resistant

Construction features

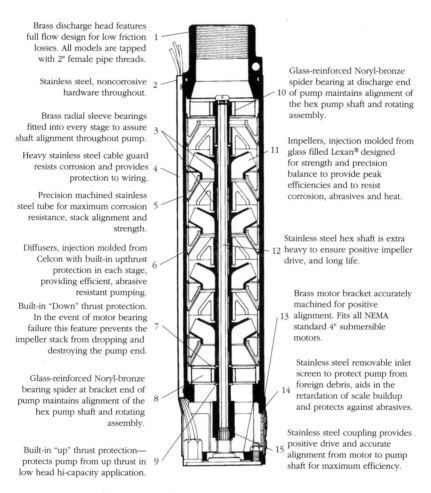

Brass discharge head features full flow design for low friction losses. All models are tapped with 2" female pipe threads. — 1

Stainless steel, noncorrosive hardware throughout. — 2

Brass radial sleeve bearings fitted into every stage to assure shaft alignment throughout pump. — 3

Heavy stainless steel cable guard resists corrosion and provides protection to wiring. — 4

Precision machined stainless steel tube for maximum corrosion resistance, stack alignment and strength. — 5

Diffusers, injection molded from Celcon with built-in upthrust protection in each stage, providing efficient, abrasive resistant pumping. — 6

Built-in "Down" thrust protection. In the event of motor bearing failure this feature prevents the impeller stack from dropping and destroying the pump end. — 7

Glass-reinforced Noryl-bronze bearing spider at bracket end of pump maintains alignment of the hex pump shaft and rotating assembly. — 8

Built-in "up" thrust protection— protects pump from up thrust in low head hi-capacity application. — 9

10 — Glass-reinforced Noryl-bronze spider bearing at discharge end of pump maintains alignment of the hex pump shaft and rotating assembly.

11 — Impellers, injection molded from glass filled Lexan® designed for strength and precision balance to provide peak efficiencies and to resist corrosion, abrasives and heat.

12 — Stainless steel hex shaft is extra heavy to ensure positive impeller drive, and long life.

13 — Brass motor bracket accurately machined for positive alignment. Fits all NEMA standard 4" submersible motors.

14 — Stainless steel removable inlet screen to protect pump from foreign debris, aids in the retardation of scale buildup and protects against abrasives.

15 — Stainless steel coupling provides positive drive and accurate alignment from motor to pump shaft for maximum efficiency.

7-9 *Cutaway of a submersible pump.* A.Y. McDonald MFG. Co., Dubuque, Iowa

be some other type. Copper tubing was used for years as a water service material. It is still used at times, but most installers now use PE pipe. When compared to copper tubing, PE pipe is less expensive, lasts longer, and is unaffected by acidic water.

PE pipe for a water service is prepared in the same way that it would be for use as a drop pipe. It is laid out and straightened to get the loops out. When this is complete, it is placed in the trench. One end of the pipe connects to the protrusion from the pitless adapter. This connection is made with two hose clamps. The other end of the pipe is placed through a foundation sleeve and extended into a home.

The entire length of the pipe should be lying flat on the bottom of the trench. There should not be any rocks or other sharp objects under or around the pipe. This is important, yet it is not something that all installers are careful about, so you might want to check it yourself. Sharp objects could puncture the pipe during the backfilling process. It is also possible for rough or sharp objects to cut the pipe months after installation when the ground settles and the pipe moves. Make sure the trench and the backfill material are free of objects that might harm the pipe.

Water service pipes often have to be inspected before they are covered. This procedure is not your direct responsibility, but you should make sure that an inspection has been approved if one is required. Backfilling the trench should be done gradually. If someone takes a backhoe and pushes large piles of dirt into the ditch, the pipe might kink or collapse. Require the backfilling to be done in layers, so that there won't be excessive weight dumped on the pipe all at once.

Inside the foundation

The rest of the work takes place inside the foundation. A pressure tank should be used on all jobs. This tank gives a reserve supply of water that allows the house to have good water pressure. The tank also preserves the life of the pump (Fig. 7-10). Without a pressure tank, the water pump would have to cut on every time someone turned on a faucet. This short cycling of the pump would wear it out quickly. A pressure tank removes the need for a pump to cut on every time water is needed. Until the tank drops to a certain pressure, the pump is not required to run. When refilling the tank, the pump runs long enough to avoid short cycling.

The size of a pressure tank is normally determined by the number of people and plumbing fixtures expected in a house. Larger tanks require the pump to run less often. In addition to the pressure tank, there are many accessories that must be installed. For example, there is a pressure gauge, a relief valve, a drain valve, a pressure switch, and a tank tee. Electrical regulations sometimes call for a disconnect box at the tank location. Even if your local code doesn't require a disconnect box, it's a good idea to have one installed.

Under normal circumstances, an experienced installation crew can complete an entire pump installation, including inside work, in

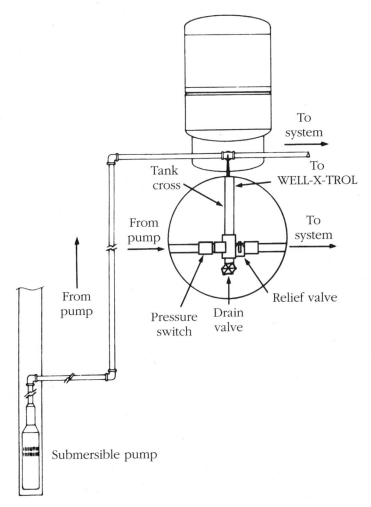

7-10 *Pressure tank in use with a submersible pump.*
Amtrol, Inc.

less than a day. As a builder, you need to hang around the job if you want to see that all of the work is being done correctly, because it happens quickly. However, you can tell a lot about the workmanship by making periodic inspections.

The finer points

We didn't discuss the finer points of pump installations, but you now have a good idea of what goes on with the process. A few other is-sues should be covered. One is pump selection. This decision is usually left up to the driller or plumber who supplies and installs a pump. It's best not to undersize a pump. While it makes no sense to buy an extremely powerful pump that is not required, it is in your customer's best interest to get a pump that exceeds its requirements by some degree. But, getting a pump that is too powerful can cause some problems. Let me explain.

Pumps are rated based on their output in gallons per minute. Sizing charts are available to show you what pumps of various sizes are capable of producing (Figs. 7-11 through 7-16). Let's say that you have a well with a recovery rate of 3 gpm. Can you imagine what might happen if you install a pump that is rated at 5 gpm. If you guessed that the well might be pumped dry, you're right. The pump should not produce more water than the well can replenish. Keep this in mind when you are looking over the specifications provided to you by bidding subcontractors. Compare the pump rate with the well rate, and make sure that the pump is not too powerful for the well.

Pumps should not be suspended too close to the bottom of a well. How far from the bottom should a pump be placed? It depends on the static water level in the well and the pumps ability to pump

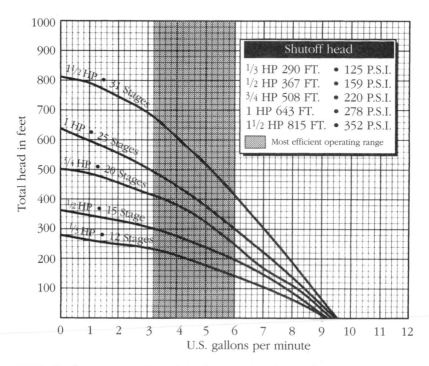

7-11 *Performance rating chart for pump with 5 gallon-per-minute output.* A.Y. McDonald MFG. Co., Dubuque, Iowa

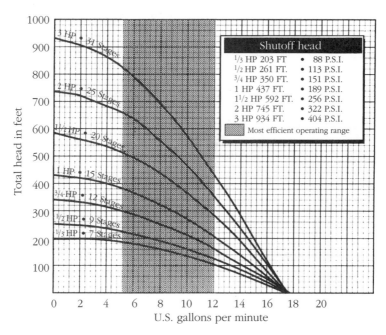

7-12 *Performance rating chart for pump with 10 gallon-per-minute output.* A.Y. McDonald MFG. Co., Dubuque, Iowa

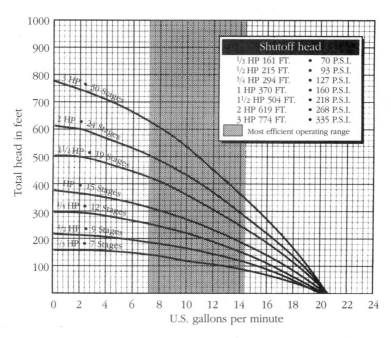

7-13 *Performance rating chart for pump with 13 gallon-per-minute output.* A.Y. McDonald MFG. Co., Dubuque, Iowa

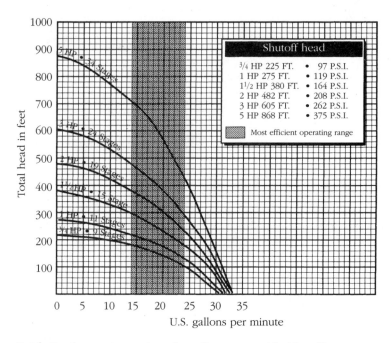

7-14 *Performance rating chart for pump with 18 gallon-per-minute output.* A.Y. McDonald MFG. Co., Dubuque, Iowa

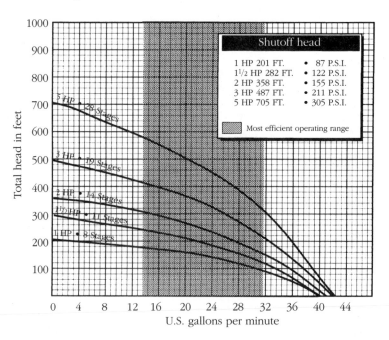

7-15 *Performance rating chart for pump with 25 gallon-per-minute output.* A.Y. McDonald MFG. Co., Dubuque, Iowa

Output in gallons per hour

Discharge pressure 0 P.S.I.

DEPTH		25'	50'	75'	100'	125'	150'	175'	200'	225'	250'	275'	300'	325'	350'	375'	400'	425'	450'	475'
J	1/3 H.P.	540	510	470	425	390	350	310	270	210	125									
	1/2 H.P.	550	520	490	465	440	410	380	350	320	290	255	220	150						
	3/4 H.P.	560	540	520	490	470	445	425	405	385	365	340	325	300	275	245	215	185	145	100
	1 H.P.	565	545	530	510	495	475	460	440	420	400	385	365	350	330	305	290	265	240	220
K	1/3 H.P.	970	890	810	725	630	530	400												
	1/2 H.P.	995	950	865	790	720	635	540	445	325	160									
	3/4 H.P.	1005	960	900	850	800	760	700	635	560	490	410	325	205						
	1 H.P.	1020	975	940	895	860	820	775	725	680	640	590	535	475	420	325	240	120		
L	1/3 H.P.	1100	970	840	680	480	205													
	1/2 H.P.	1120	1025	930	820	710	570	420	210											
	3/4 H.P.	1150	1085	1020	955	875	800	715	620	505	370	200								
	1 H.P.	1170	1115	1060	1005	950	880	820	755	685	605	520	420	300	155					
P	3/4 H.P.	1800	1650	1560	1440	1290	1110	870	600											
	1 H.P.	1830	1725	1650	1560	1440	1305	1140	960	720	450									
M	1 H.P.	2290	2110	1930	1740	1500	1152	720												

Discharge pressure 30 P.S.I.

DEPTH		25'	50'	75'	100'	125'	150'	175'	200'	225'	250'	275'	300'	325'	350'	375'	400'	425'	450'	475'
J	1/3 H.P.	440	395	360	320	275	230	160												
	1/2 H.P.	475	445	415	385	360	330	295	260	220	175									
	3/4 H.P.	505	475	450	425	410	390	370	350	335	305	280	250	230	190	150	105			
	1 H.P.	515	500	480	460	450	425	400	385	370	355	330	310	295	270	245	220	205	175	145
K	1/3 H.P.	785	695	603	488	333														
	1/2 H.P.	815	730	670	550	470	360													
	3/4 H.P.	875	815	755	705	660	590	505	420	360	240									
	1 H.P.	910	865	820	780	745	695	650	600	555	480	420	335	275						
L	1/3 H.P.	708	495																	
	1/2 H.P.	795	730	595	440															
	3/4 H.P.	985	890	815	733	635	525	393												
	1 H.P.	1015	958	895	833	765	695	608	535	433	320									
P	3/4 H.P.	1470	1320	1140	920	600														
	1 H.P.	1570	1460	1230	1180	950	760													
M	1 H.P.	1778	1540	1194	785															

Discharge pressure 40 P.S.I.

DEPTH		25'	50'	75'	100'	125'	150'	175'	200'	225'	250'	275'	300'	325'	350'	375'	400'	425'	450'	475'
J	1/3 H.P.	395	365	325	280	215	170	120												
	1/2 H.P.	440	420	390	360	330	300	280	230	180	105									
	3/4 H.P.	480	455	430	410	385	370	355	330	310	285	255	225	195	155	120				
	1 H.P.	500	485	465	445	425	405	390	370	355	335	310	295	265	250	225	205	175	145	115
K	1/3 H.P.	745	650	555	430	240														
	1/2 H.P.	745	670	615	485	365	230													
	3/4 H.P.	825	780	720	660	590	515	445	360	250										
	1 H.P.	875	830	790	745	695	660	585	550	495	440	360	275	170						
L	1/3 H.P.	550	280																	
	1/2 H.P.	745	620	470	280															
	3/4 H.P.	900	820	740	650	540	415	260												
	1 H.P.	960	900	835	775	700	630	520	455	335	205									
P	3/4 H.P.	1320	1230	920	690															
	1 H.P.	1460	1380	1200	1020	790														
M	1 H.P.	1588	1272	852	360															

Discharge pressure 50 P.S.I.

DEPTH		25'	50'	75'	100'	125'	150'	175'	200'	225'	250'	275'	300'	325'	350'	375'	400'	425'	450'	475'
J	1/3 H.P.	365	325	280	235	180														
	1/2 H.P.	420	390	360	330	300	265	230	175	110										
	3/4 H.P.	460	435	415	390	370	355	330	310	285	260	225	200	160	115					
	1 H.P.	485	470	445	425	405	390	370	355	335	315	295	270	250	230	205	175	145	130	
K	1/3 H.P.	605	475																	
	1/2 H.P.	670	585	480	360	240														
	3/4 H.P.	770	720	660	585	515	445	360	250	120										
	1 H.P.	835	790	745	695	660	610	550	490	430	365	275	170							
L	1/2 H.P.	618	448																	
	3/4 H.P.	830	743	653	545	410														
	1 H.P.	905	843	778	708	630	550	435	340											
P	3/4 H.P.	1170	960	720																
	1 H.P.	1390	1200	1030	810															
M	1 H.P.	1232	876																	

FRICTION LOSSES IN RISER PIPE HAVE NOT BEEN CALCULATED

7-16 *Output performance chart for submersible pump.* A.Y. McDonald MFG. Co., Dubuque, Iowa

high enough. A pump placed too close to the bottom of a well can pick up gravel, sand, and other debris. It might not have enough power to pump water to a pressure tank. Check performance charts before you buy a pump. Some wells fill in a little over time. Having a pump too close to the bottom when a well is filling is bad news. A pump should be, in my opinion, at least 15 feet above the bottom, and higher when practical.

My personal pump is about 30 feet above the bottom of the well. The depth is a little over 400 feet. The static level is only about 15 feet from the top of my well casing. This means that I have a column of water that is about 385 feet deep. That's a substantial reserve. I would have to pump hundreds of gallons of water at one time to run the well dry. Since I have so much water, it's easy for me to keep my pump hanging high above the bottom. Not all wells have so much reserve water, and this can force an installer to hang a pump closer to the bottom. The key is to keep the pump far enough from the bottom to keep debris from getting sucked into the pump. If you want to move a lot of water, choose a pump with a high gpm rating, assuming that the well has a strong recovery rate.

Problems

Problems sometimes come up with well installations. We are going to talk about most of them in Chapter 16. There is one problem, however, that we should discuss at this point. What would you do if the trench for your water service could not be dug deep enough to protect the pipe from freezing? This doesn't happen often, but it does happen. In fact, it happened at my house. Bedrock was between 18 and 24 inches down, so getting to a depth of four feet was not feasible. Winters are very cold in Maine, so there was no doubt that my water service would freeze if left unprotected at such a shallow depth. You might run into a similar situation.

In my case, I took three precautions. I insulated the water service pipe with foam pipe insulation. Then I ran it through a continuous sleeve of 3-inch plastic pipe. The air space between the sleeve and the water service adds to frost protection. My final, and most effective step was to install an in-line heat tape. This type of heat tape is expensive, but it's a bargain when you are faced with difficult situations, such as mine.

The heat tape is waterproof. It runs the full length of the water pipe. The heating element is inside the water pipe, in the water. A sensor and control is installed in some practical location. Mine is in a crawl space, with the temperature sensor positioned by an air vent. This allows the sensor to feel the outside temperature.

During the winter, I plug in the control unit, which is thermostatically controlled, and the heat tape runs automatically. It heats the water in the pipe to a point above freezing. I used it all last winter without any problems. If you can't dig as deep as you would like for the water service, I strongly recommend an in-line heat tape.

We are now ready to move along to Chapter 8, where we are going to discuss dug wells and their pump systems.

8

Shallow wells

Shallow wells are very different from drilled wells. While most drilled wells have diameters of six inches, a typical shallow well has a diameter of about three feet. Concrete usually surrounds a shallow well, rather than the steel casing used for a drilled well. Drilled wells often reach depths of 300 feet, but a shallow well rarely runs deeper than 30 feet. There are, to be sure, many differences between the two types of wells.

Shallow wells are much less expensive to install than drilled wells. These wells go by many names. I call them dug wells. Sometimes they are called bored wells. Without trying to get extremely technical, I am going to simply refer to these wells as dug wells. You now know that I might be lumping bored wells into the category and that both types are considered shallow wells, so we shouldn't have a problem understanding each other.

In the old days, dug wells were created with picks, shovels, and buckets. The work was dangerous. Today, boring equipment is normally used to create a dug well. To be specific about the type of well I'm talking about, let me explain the basic makeup of one.

If you see a concrete cylinder that has a diameter of approximately three feet sticking up out of the ground, you are looking at what I call a dug well. The concrete casing typically is covered with a large, heavy concrete disk.

Old dug wells were dug by hand. Many of them were lined with stones. I've crawled down a few of these as a plumber, to work on pipes and such, but I don't think I'd do it today. My younger years as a plumber were more adventuresome than I would care to repeat.

Dug wells are shallow when compared to drilled wells. Finding a dug well that is 50 feet deep would be similar to finding pirate treasure sitting on top of a sandy beach. Most dug wells I've seen have been no more than 35 feet deep. That's why jet pumps, which are often used with these type wells, are called shallow-well pumps.

Bored wells can run much deeper. It's possible for a bored well to reach a depth of 100 feet or more. A two-pipe jet pump or a submersible pump is needed for a deep well (Fig. 8-1). The depth is regulated by the ground into which the well is dug. Some types of earth allow for a deeper well than others. Still, in practical terms, most dug or bored wells won't exceed 50 feet in depth. Beyond this level, a drilled well is more practical. A small-diameter deep well can be pumped with a single-pipe pump (Fig. 8-2).

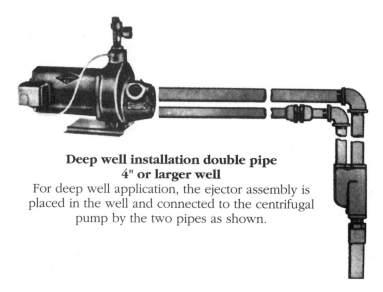

**Deep well installation double pipe
4" or larger well**
For deep well application, the ejector assembly is placed in the well and connected to the centrifugal pump by the two pipes as shown.

8-1 *Two-pipe pump system for a deep well.*
A.Y. McDonald MFG. Co., Dubuque, Iowa

Due to the large diameter of a dug well, a lot of water can be stored in reserve. While the water in a shallow well might be only a few feet deep, it has a lot more surface area than water stored in a deep well. The increased surface area helps to make up for the lack of depth.

Shallow wells are common in many parts of the country. When the water table is high and reasonably constant, shallow wells work fairly well. They might dry up during some hot, dry times of the year, but this is not always the case. Some shallow wells maintain a good volume of water all through the year.

If they don't run completely dry, they might contain such a small quantity of water that rationing is needed. For example, you might have to go to a local laundromat to wash clothes for a few weeks out of the year. This, of course, is more inconvenience than some home-owners are willing to put up with.

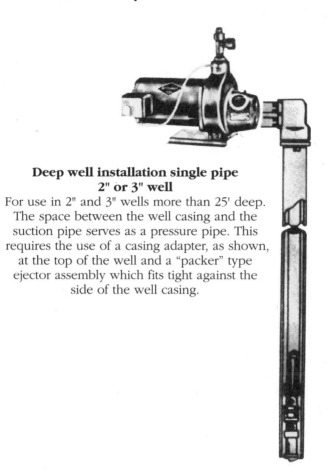

**Deep well installation single pipe
2" or 3" well**
For use in 2" and 3" wells more than 25' deep.
The space between the well casing and the
suction pipe serves as a pressure pipe. This
requires the use of a casing adapter, as shown,
at the top of the well and a "packer" type
ejector assembly which fits tight against the
side of the well casing.

8-2 *One-pipe pump system for a deep well that has a
small diameter.* A.Y. McDonald MFG. Co., Dubuque, Iowa

Most of the homes that I've owned have had shallow wells. My
present home has a drilled well, and I feel much more secure about
it than I ever did with shallow wells. But, only once did I ever have
trouble with one of my shallow wells, and that trouble occurred in
the summer.

When should
you use a shallow well?

When should you use a shallow well? The ground conditions at the
well site have much to do with the decision. If bedrock is near the
ground surface, a dug well is not practical. Drilled wells are the an-

swer when you have to penetrate hard rock. To give you a good idea of what depths are possible and in what types of soil conditions different types of shallow wells can be used, let me break them down into categories.

True dug wells

True dug wells can be created to depths up to 50 feet. The diameters can range from 3 feet, the most common size, to a massive 20 feet. True dug wells can be created in clay, silt, sand, gravel, and cemented gravel. You can even deal with boulders. You might be able to build a true dug well in sandstone or limestone, but the material must either be soft or fractured. Dense igneous rock can not be penetrated with a dug well.

Bored wells

Bored wells can reach depths of 100 feet. The diameter of this type of well can be as small as 2 inches. Larger diameters, in the 30-inch range, are also common. Clay, silt, sand, and gravel can all support a bored well. Cemented gravel can stop a bored well, and so can bedrock. You can sometimes work around boulders. Sandstone and limestone affect a bored well in the same way that they affect a dug well.

Drilled wells

Drilled wells can run to depths of 1,000 feet. This is far from being a shallow well. All of the geologic formations mentioned for the well types above can be overcome with a drilled well. Percussion and hydraulic rotary drilling are very similar in their abilities. Well seals (Fig. 8-3) are used to protect the openings of drilled wells while allowing one or two well pipes to enter the casing.

Jetted wells

Jetted wells can run up to 100 feet in depth. They cannot be installed in bedrock, limestone, or sandstone. Boulders, cemented gravel, and large loose gravel all prohibit the use of jetted wells. Jetted wells are best suited for use in clay, sand, and silt. Diameters range from 2 to 12 inches.

Characteristics

Let's talk about some of the characteristics of shallow wells. Dug wells can extend only a few feet below a water table. The geologic condi-

AW5 **AW6**

AW5 well seals for twin pipe installations:
The sanitary well seals prevents foreign
matter from getting in the well. Opening for
vent pipe.

8-3 *Well seals used to close the opening of
a drilled well.* A.Y. McDonald MFG. Co., Dubuque, Iowa

tions affect the recovery rate of a well. For example, a dug well that is
surrounded by gravel might produce an excellent recovery rate, while
one having fine sand surrounding it might produce a poor rate of recovery. Dug wells cannot be considered terrific water producers.

Bored wells are a little different. They can extend deeper into the
water table. It is not uncommon to go ten feet below the edge of a
water table. This added depth gives bored wells an advantage over
dug wells. Having an 8-foot head of water with a 3-foot diameter provides a good supply of water, assuming that the recovery rate is good.
It would not be considered strange or unusual for a bored well to
have a recovery rate of 20 gpm, about double what would be expected of a good dug well.

Even though many professionals, myself included, call all shallow
wells dug wells, there is clearly a difference between a bored well
and a true dug well. Most shallow wells that are installed today are
bored wells. If your well installer talks about a dug well, make sure
that the installer is actually referring to a bored well.

Some bored wells hit artesian aquifers. When this happens, the
static water level in the well rises. For example, a bored well that extends ten feet past the water table would normally be thought of as
holding ten feet of water. If an artesian aquifer is hit, the actual water
reserve could be twenty feet deep, or more. I've seen bored wells
with such powerful artesian effects that water actually ran out over
the top of the well casing. It is certainly possible to obtain plenty of
water for a residence with a bored well. But, there is no guarantee.

Look around

If you are unsure of what type of well to use, look at the wells that
are being used for houses near your building site. If you see a lot of

concrete casings sticking out of the ground, it's a good indication that shallow wells work in the area. A host of steel casings indicate a need for a drilled well. By looking at neighboring homes, you can get a good feel for the type of well that is likely to be needed. It is not, however, a sure bet.

If you want more dependable information on the type of well that is likely to be suitable for your project, talk to some experts. Well installers are an excellent place to start. Local installers should know a great deal about the pros and cons of various types of wells in the region where you plan to build. But, be careful. You might run across an installer who is not equipped to install all types of wells. If this is the case, it's possible that the recommendations might be in the best interest of the installer rather than you or your customer.

Most rural areas have county extension offices or some similar facility to help with all sorts of issues. Wells fall into this category. A visit to the right local agency might reveal detailed maps that show aquifers and soil types. If you can pinpoint your building location on a map of this type, you can make some sound decisions on the type of well to use.

Combined research is the true key to success. Visit your county extension office and look over the maps. Take an inventory of wells on surrounding properties. Talk to several well installers. Combine all the knowledge and advice you collect to make a wise decision. Don't be afraid to install a bored well if evidence points to it as a good choice.

As a builder in Virginia, I was responsible for the installation of dozens upon dozens of shallow wells. To the best of my knowledge, only one of the wells ever gave anyone any trouble, and that was one on my personal property. When the conditions are right and the job is done properly, a shallow well can give years of quality service at an affordable price.

The preliminary work

The preliminary work for a builder who is having a shallow well installed is very similar to that of a builder who is preparing for the installation of a drilled well. Since we covered the steps for this in the last chapter, I won't repeat them here. The only big difference is the size of a shallow well. A drilled well can be hidden easily with a few strategically placed shrubs. The casing and cap of a shallow well is much harder to hide. Aside from the size of the well, the rest of the steps are about the same.

After a well is in

After a well is installed, you are faced with the disinfection process. Just as we discussed in the last chapter, chlorine is normally used to purify a well before the water is tested. However, a few extra things must be considered when testing a shallow well. For one thing, the well cover is heavy. One person can slide the cover aside, but the job is easier with two people. The concrete cover is brittle. Rough handling can cause it to crack or break. Having a concrete cover drop and catch your finger between the lip of the well casing and the cover could be extremely painful. Be careful when handling the cover.

The odds of a child or an animal falling into an uncovered drilled well are much lower than they are with a shallow well. Two adults could jump into the opening of a bored well without any problem. Never leave the cover off of a shallow well when you are not right at the well site. Even if you are just going into the house to turn the water on or off, put the cover back on the well. Pets and people can fall into an open well quickly, and the results can be disastrous. Even drilled wells should be covered every time they are left unattended. Curious kids and pets can find themselves in some very dangerous circumstances in the blink of an eye.

You must monitor the water level closely when purging chlorine from a shallow well. While it is uncommon for a drilled well to be pumped dry during a purging, it is not unusual for a dug or bored well to have it's water level fall below the foot valve or end of the drop pipe. If this happens, the pump keeps pumping, but it's only getting air. This is bad for the pump and can burn it up. Monitor the water level closely as you empty a shallow well.

You can watch the water level in most shallow wells without any special equipment. Your eyes should be the only tools needed to tell when the water level drops down too far. However, a string with a weight attached to it can be used to gauge the water depth. Make the string a foot shorter than the drop pipe. If you don't know the length of the drop pipe, your pump installer should be able to tell you. As long as the string hits water each time it is dropped into the well, you know the water level is safe.

Pump selection

Selecting a pump for a shallow well is sometimes a little more complex than choosing one for a deep well. For example, you would probably use a jet pump, but not a multi-stage unit (Fig. 8-4). Almost

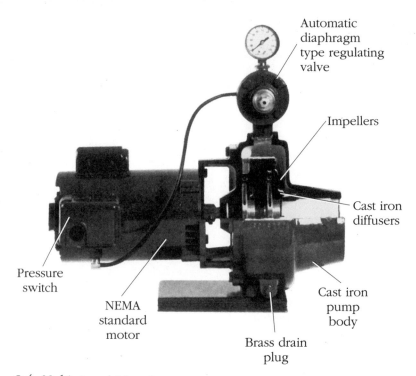

Automatic
diaphragm
type regulating
valve

Impellers

Cast iron
diffusers

Pressure
switch

NEMA
standard
motor

Cast iron
pump
body

Brass drain
plug

8-4 *Multi-stage jet pump.* A.Y. McDonald MFG. Co., Dubuque, Iowa

everyone uses submersible pumps for drilled wells. It's possible to use a two-pipe jet pump for a deep well, but few installers recommend this course of action. Shallow wells, on the other hand, can use single-pipe jet pumps (Fig. 8-5), two-pipe jet pumps, or even submersible pumps. But, as a rule, multi-stage pumps (Fig. 8-6) are not used in shallow wells.

Performance charts (Fig. 8-7) can guide you to the proper pump. The use of submersible pumps and multi-stage pumps (Fig. 8-8) in shallow wells is not common. In wells where the water lift is not more than about 25 feet, single-pipe jet pumps (Fig. 8-9) are often used. Deeper bored wells, and some dug wells, see the use of two-pipe jet pumps. If a well of any type has an adequate depth of water, a submersible pump can always be used.

Single-pipe jet pumps

Single-pipe jet pumps (Fig. 8-10) are the least expensive option available for a shallow well. But, their lifting capacity is limited (Fig. 8-11). Since single-pipe jet pumps work on a suction-only basis, they must be able to pull a vacuum on the water. Since physics plays a part in

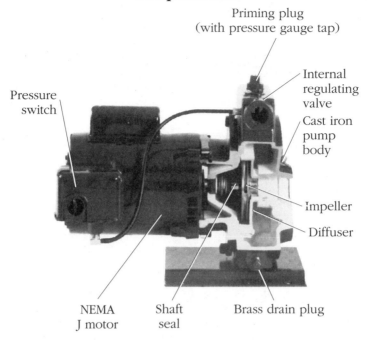

Priming plug
(with pressure gauge tap)

Internal
regulating
valve

Pressure
switch

Cast iron
pump
body

Impeller

Diffuser

NEMA
J motor

Shaft
seal

Brass drain plug

8-5 *Convertible jet pump.* A.Y. McDonald MFG. Co., Dubuque, Iowa

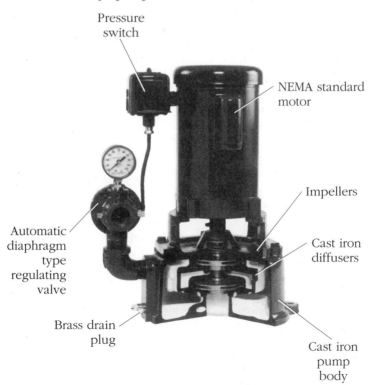

Pressure
switch

NEMA standard
motor

Impellers

Cast iron
diffusers

Automatic
diaphragm
type
regulating
valve

Brass drain
plug

Cast iron
pump
body

8-6 *Multi-stage jet pump.* A.Y. McDonald MFG. Co., Dubuque, Iowa

8100 Series Specifications ▼ Shallow Well

MODEL NO.	HP	VOLTS	IMPELLER MATERIAL
8130	1/3	115	Plastic
8131	1/3	115	Brass
8150	1/2	115/230	Plastic
8151	1/2	115/230	Brass
8170	3/4	115/230	Plastic
8171	3/4	115/230	Brass
8110	1	115/230	Plastic
8111	1	115/230	Brass

8-7 *Performance ratings for jet pumps.* A.Y. McDonald MFG. Co., Dubuque, Iowa

Multi-Stage Specifications
▼ 1500 Series Horizontal ▼ 1000 Series Vertical

MODEL	MODEL	HP	VOLTS	IMPELLER MATERIAL
1550	1050	1/2	115/230	Brass
1575	1075	3/4	115/230	Brass
1575SW	1075SW	3/4	115/230	Brass
1510	1010	1	115/230	Brass
1510SW	1010SW	1	115/230	Brass
1515SW	1015SW	1 1/2	115/230	Brass

8-8 *Performance ratings for multi-stage pumps.* A.Y. McDonald MFG. Co., Dubuque, Iowa

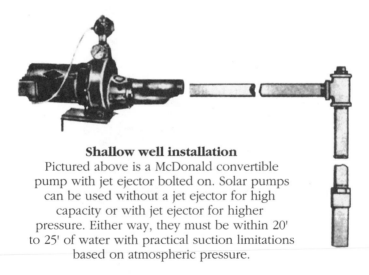

Shallow well installation
Pictured above is a McDonald convertible pump with jet ejector bolted on. Solar pumps can be used without a jet ejector for high capacity or with jet ejector for higher pressure. Either way, they must be within 20' to 25' of water with practical suction limitations based on atmospheric pressure.

8-9 *Single-pipe, shallow-well pump.* A.Y. McDonald MFG. Co., Dubuque, Iowa

PRES. SWITCH SETTING	SUCTION PIPE SIZE	DISCHARGE SIZE	SHIPPING WEIGHT
20-40	1 1/4"	3/4"	46 lbs.
20-40	1 1/4"	3/4"	48 lbs.
20-40	1 1/4"	3/4"	48 lbs.
20-40	1 1/4"	3/4"	50 lbs.
30-50	1 1/4"	3/4"	50 lbs.
30-50	1 1/4"	3/4"	52 lbs.
30-50	1 1/4"	3/4"	52 lbs.
30-50	1 1/4"	3/4"	53 lbs.

8-7 *Continued.*

PRES. SWITCH SETTING	SUCTION PIPE SIZE	TWIN TYPE DROP PIPE	SHIPPING WEIGHT
30-50	1 1/4"	1" x 1 1/4"	65 lbs.
30-50	1 1/4"	1" x 1 1/4"	71 lbs.
30-50	1 1/4"	1" x 1 1/4"	66 lbs.
30-50	1 1/4"	1" x 1 1/4"	74 lbs.
30-50	1 1/4"	1" x 1 1/4"	67 lbs.
30-50	1 1/4"	1" x 1 1/4"	72 lbs.

8-8 *Continued.*

how high above sea level a vacuum can be made, jet pumps can only pull water so high. The exact maximum lift under ideal conditions is around 30 feet. But, for practical purposes, most professionals agree that a suction-based pump should not be expected to lift water more than 25 feet.

Jet pumps are installed at some location outside of the well. Unlike submersible pumps, which hang in the well water, jet pumps are normally installed in basements, crawl spaces, or pump houses (Fig. 8-12). The pumps and their pipes must be protected from freezing temperatures. A pressure tank should be used with a jet pump. Many jet pumps are made to sit right on top of a pressure tank (Fig. 8-13), with the use of a special bracket. The pressure tank must also be protected from freezing temperatures.

Two-pipe jet pumps

Two-pipe jet pumps can be used to pump water from deeper wells. These pumps use two pipes. The pump forces water down one pipe

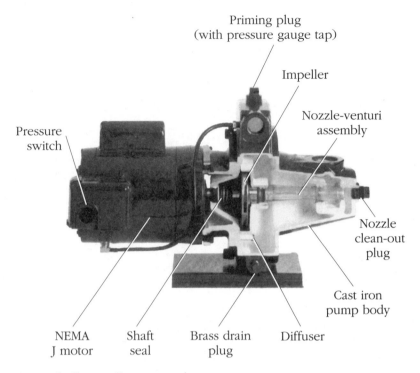

Priming plug
(with pressure gauge tap)

Impeller

Nozzle-venturi
assembly

Pressure
switch

Nozzle
clean-out
plug

Cast iron
pump body

NEMA   Shaft   Brass drain   Diffuser
J motor   seal   plug

8-10 *Shallow-well jet pump.* A.Y. McDonald MFG. Co., Dubuque, Iowa

8100 Series ▼ High Capacity ▼ Shallow Well

MODEL NO.	HP	SUCTION LIFT - FT.	20
8130 **8131**	1/3	5	755
		15	600
		25	345
8150 **8151**	1/2	5	935
		15	700
		25	395
8170 **8171**	3/4	5	1130
		15	865
		25	515
8110 **8111**	1	5	1605
		15	1240
		25	775

8-11 *Shallow-well performance chart.* A.Y. McDonald MFG. Co., Dubuque, Iowa

to allow it to be sucked up in the other (Fig. 8-14). Since a two-pipe pump is not dependent solely on suction, it can handle higher lifts. The physical appearance of a two-pipe pump is basically the same as that of a one-pipe pump, except for the extra pipe.

Submersible pumps

Submersible pumps push water up out of a well. Since the pump is pushing, rather than sucking, it can be installed at great depths. In many ways, submersible pumps are far less prone to failure than jet pumps, probably because there are fewer parts to fail. Although more expensive, they tend to last longer. This makes submersible pumps a favorable choice when the water depth is sufficient to warrant their use.

Since jet pumps are most often used with shallow wells, we are going to concentrate our efforts on them. Submersible systems were described in the previous chapter. Since most shallow wells are equipped with a single-pipe jet pump, we are going to start our installation procedures with them.

Installing a single-pipe jet pump

It's fairly easy to install a single-pipe jet pump (Fig. 8-15). The well portion of the work is particularly simple. Almost anyone with mod-

CAPACITIES IN GPH AT DISCHARGE PRESSURE		
30	**40**	**50**
755	445	150
600	360	80
345	120	--
935	515	135
700	390	70
395	280	--
1130	970	525
865	740	235
515	440	160
1605	1545	1385
1240	1210	620
775	760	440

8-11 *Continued.*

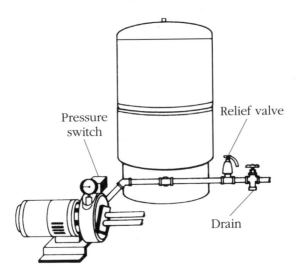

Pressure
switch

Relief valve

Drain

8-12 *A typical jet-pump set-up.* Amtrol, Inc.

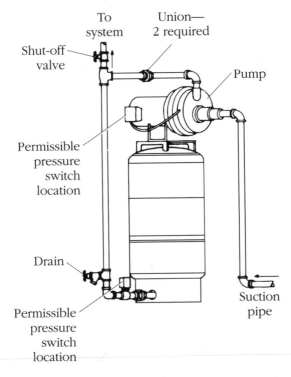

To
system

Union—
2 required

Shut-off
valve

Pump

Permissible
pressure
switch
location

Drain

Suction
pipe

Permissible
pressure
switch
location

8-13 *A jet pump mounted on a pressure tank with a pump bracket.* Amtrol, Inc.

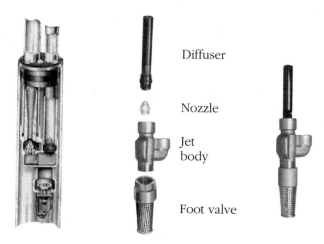

8-14 *In-well assembly system for a two-pipe pump.*
Goulds Pumps, Inc.

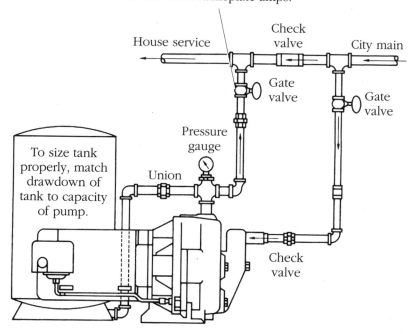

8-15 *A typical piping arrangement for a jet pump.* Goulds Pumps, Inc.

est mechanical skills and an ability to read instructions can manage the installation of a jet pump (Fig. 8-16). However, you probably need a license to do the work. This might prohibit you from making your own installations.

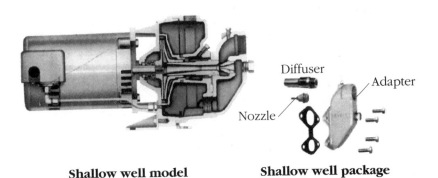

Shallow well model   **Shallow well package**

8-16 *Cross section of a jet pump.* Goulds Pumps, Inc.

The first step in the installation of a jet pump is to dig a trench between the well casing and the pump location. Since we covered trenching in the last chapter, we can skirt the issue here. Once the trench is dug, you are ready to make a hole in the side of the well casing. Since most shallow wells are cased with concrete, a cold chisel and a heavy hammer are needed to make a hole. The hole must be patched to make it watertight after the well piping is installed. Otherwise, groundwater can run into the well, possibly causing contamination.

After a hole is made, you are ready to install the well pipe. The pipe is probably going to be PE pipe. Insert fittings should be made of metal and all connections should be double clamped. A foot valve (Fig. 8-17) is usually installed on the end of the pipe that hangs in the well water. The foot valve serves two purposes. It acts as a strainer to block out gravel and similar debris that might otherwise be sucked into the pipe and pump. Foot valves also act as check valves, to prevent water in the drop pipe from running out into the well. If the water in a drop pipe was not controlled by a check valve (Fig. 8-18) or foot valve, the jet pump would lose its prime and fail to pump water.

After the foot valve is installed, an elbow fitting is attached to the other end of the pipe. This fitting allows the water service pipe, which is going to be placed in the trench, to connect to the drop pipe. Some form of protection should be provided for the water service pipe where it penetrates the concrete casing. Foam pipe insulation can sometimes be used, and rigid plastic pipe, like the type used for

AV5 foot valves: Bronze body with stainless steel strainer.

Order no.	Pipe size	Outside diameter
AV5-07	3/4"	1^{11}/$_{16}$"
AV5-10	1"	1^{15}/$_{16}$"
AV5-12	1^1/$_4$"	2^1/$_4$"
AV5-15	1^1/$_2$"	2^7/$_8$"
AV5-20	2"	3^1/$_2$"

8-17 *Foot valve.* Goulds Pumps, Inc.

AV6 check valves: Bronze construction.

Order no.	Pipe size
AV6-10	1"
AV6-12	1^1/$_4$"
AV6-15	1^1/$_2$"
AV6-20	2"
AV6-30 (iron)	3"
AV6-40 (iron)	4"

8-18 *Check valve.* Goulds Pumps, Inc.

drains and vents in plumbing systems, can always be used. If PE pipe is allowed to rest on the rough edge of concrete, the constant vibration of the pipe when the pump is operating can eventually wear a hole through the pipe.

After the connection is made between the drop pipe in the well and the water service pipe in the trench, the rest of the work is done at the pump location. Backfilling the trench should be done with the same precautions described in the previous chapter.

Jet pumps can be installed almost anywhere where they are not going to get wet or become frozen. Crawl spaces, basements, pump houses, and even closets are all potential installation locations. A pressure tank is usually installed in close proximity to a jet pump. In some cases, the pump is bolted to a bracket on a pressure tank (Fig. 8-19).

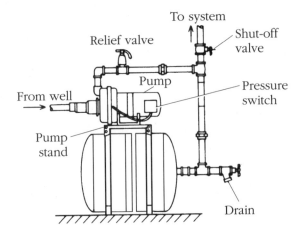

8-19 *Bracket-mounted jet pump on a horizontal pressure tank.* Amtrol, Inc.

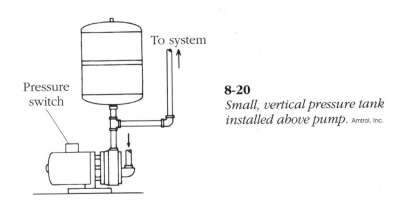

8-20

Small, vertical pressure tank installed above pump. Amtrol, Inc.

Sometimes a small pressure tank is suspended above a pump (Fig. 8-20 and 8-21). Many tanks stand independently (Fig. 8-22) on a solid surface, such as a floor.

Sometimes the decision regarding the type of tank to be used, and where it should be located, is at least partly based on the space that's available for the tank and pump. It is not mandatory to use a pressure tank, but it is highly recommended in order to prolong the life of a pump. However, a pressure tank does not have to be installed adjacent to a pump. It can be in a remote location (Fig. 8-23). Pressure tanks can even be installed underground, as long as the tank is an underground model (Fig. 8-24).

The piping arrangement for a jet pump is not complicated. However, many common accessories are used, such as an air-volume con-

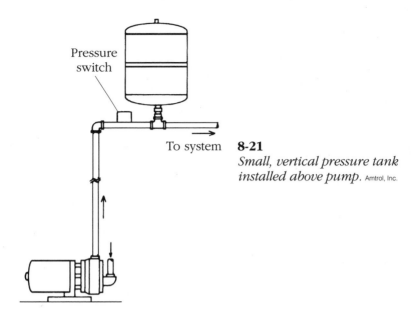

Pressure switch

To system **8-21**
*Small, vertical pressure tank
installed above pump.* Amtrol, Inc.

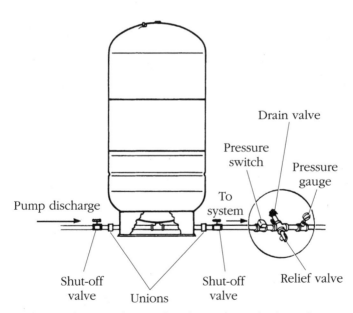

Drain valve

Pressure
switch

Pressure
gauge

Pump discharge

To
system

Shut-off
valve   Shut-off   Relief valve
      Unions   valve

Note: When replacing galvanized tanks, use straight-through connection.

8-22 *Stand-type pressure tank installed with a straight-through method
not using a tank tee.* Amtrol, Inc.

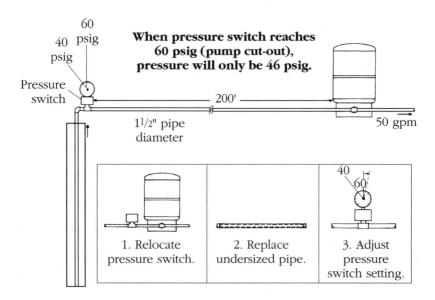

8-23 *Sizing suggestions for unusual situations.* Amtrol, Inc.

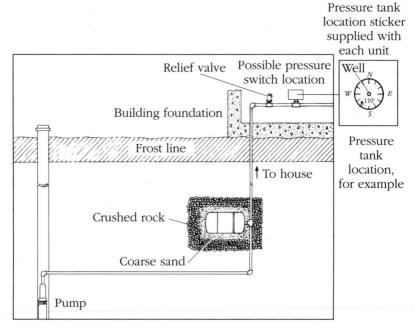

8-24 *An underground installation of a pressure tank.* Amtrol, Inc.

AA8-7 Air-volume control for jet pumps:
Diaphragm type air-volume control for mounting
on side of tank. Plastic spring insert protects against scratching.
Threaded 3/8" male for tank connection. 1/8" × 1/4" connector
for 1/4" copper tubing. No tubing or fittings
included.

8-25 *Air-volume control for a jet pump.* Goulds Pumps, Inc.

AA1 Air-volume control for series BF pumps:
The AA1 air-volume control admits
air when needed and automatically maintains
the proper volume. Control body
threaded 11/4" pipe size. Weight 1 lb. Tubing
not included.

8-26 *Float-type air-volume control.*
Goulds Pumps, Inc.

8-27
Pressure switch set-up.
Goulds Pumps, Inc.

trol (Figs. 8-25 and 8-26). Shut-off valves are, of course, installed in the pipes that leave the pump. In the case of a one-pipe jet pump, one large pipe brings water to the pump. A smaller pipe distributes the water to a water distribution system. A pressure switch (Fig. 8-27)

8-28
Relief valve. Goulds Pumps, Inc.

8-29
Pressure-control valve.
Goulds Pumps, Inc.

controls when the pump cuts on and off. Relief valves (Fig. 8-28) are needed to protect pressure tanks. Drain valves are typically installed for pressure tanks, and a check valve might be installed. In special situations, a pressure-control valve (Fig. 8-29) might be desirable.

Two-pipe pumps

Two-pipe pumps look very similar to one-pipe pumps, except for the extra pipe involved. While one-pipe jet pumps have only a suction pipe, two-pipe pumps have both a pressure and a suction pipe. Some differences in the well piping arrangement go along with this most noticeable difference. An ejector is installed to enable the pressure pipe to assist the suction pipe in lifting water from the well.

A pressure tank should still be used with a two-pipe system. In fact, a pressure tank should be installed with all types of well pumps for residential use. Let's spend some time talking about pressure tanks in detail.

Pressure tanks

Pressure tanks (Fig. 8-30) should be considered standard equipment with every residential pump installation. The tanks are not very expensive and they can add years to the life of a pump. They can also provide residents with better water pressure, an important aspect to

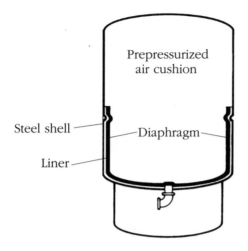

Prepressurized
air cushion

Steel shell

Diaphragm

Liner

8-30 *Diaphragm-type pressure tank.*
Goulds Pumps, Inc.

consider. Today, it's rare to find a well plumbing system that doesn't include a pressure tank. I can't imagine a pump installer bidding a job without including the cost of a pressure tank, but make sure that your next well system does provide a pressure tank. With that said, let's talk specifically about the various sizes and types of pressure tanks that are available.

Small tanks

Small tanks are available in sizes that hold no more than two gallons of water (Fig. 8-31). This is a very small tank. In my opinion, it is too small for almost any application. A tank that has such a minimum capacity does almost no good in a routine, residential plumbing system. When you consider that many toilets use more than two gallons of water each time they flush, you can see that the pump is going to run a lot. Think about showers. At a flow rate of three gallons per minute, the first minute of a shower depletes the supply of a small pressure tank.

The purpose of a pressure tank is to take some strain off a pump. While a two-gallon buffer provides some support to a pump, the help is minimal. A pressure tank should be sized to meet the needs of the house it serves. In other words, the number of people using the plumbing system and the types of fixtures available should be taken into consideration when choosing a pressure tank.

In-line models

Model No.	Dimensions Diameter (ins)	Height (ins)	Total Volume (gals)	Max. Accept. Factor	Drawdown 20/40 (gals)	30/50 (gals)	40/60 (gals)	Shipping Wt.(Vol.) lbs(cu ft)
WX-101	8	12 5⁄8	2.0	0.45	.7	.6	.5	5 (0.6)

8-31 *In-line pressure tank.* Amtrol, Inc.

Large tanks

Large tanks (Fig. 8-32) can take up a substantial amount of room. This is not normally a problem in houses where a basement, cellar, or crawl space is available. But, if the pump system and tank must be installed within the primary living space of a home, tank size can become an issue. Residential pressure tanks can be purchased with an ability to hold more than 100 gallons of water. In most circumstances, this is overkill. A tank that holds between twenty and forty gallons should perform well under average residential conditions.

Stand models

Model No.	Dimensions Diameter (ins)	Height (ins)	Total Volume (gals)	Max. Accept. Factor	Drawdown 20/40 (gals)	30/50 (gals)	40/60 (gals)	Shipping Wt. (Vol.) lbs (cu ft)
WX-104-S	15 3⁄8	19 1⁄4	10.3	1.00	3.8	3.2	2.8	25 (3.0)
WX-201	15 3⁄8	23 7⁄8	14.0	0.81	5.1	4.3	3.7	27 (3.8)
WX-202	15 3⁄8	31 5⁄8	20.0	0.57	7.3	6.2	5.4	35 (4.9)
WX-203	15 3⁄8	46 3⁄8	32.0	0.35	—	9.9	8.6	43 (7.0)
WX-104-LTD	15 3⁄8	19 1⁄4	10.3	1.00	3.8	3.2	2.7	23 (3.0)
WX-201-LTD	15 3⁄8	24	14.0	0.81	5.2	4.3	3.8	25 (3.8)
WX-202-LTD	15 3⁄8	31 3⁄4	20.0	0.57	7.4	6.2	5.4	33 (4.9)
WX-203-LTD	15 3⁄8	46 5⁄8	35.0	0.35	—	9.9	8.6	43 (7.0)
WX-205	22	29 1⁄2	34.0	1.00	12.4	10.5	9.1	61 (9.5)
WX-250	22	35 5⁄8	44.0	0.77	16.3	13.6	11.9	69 (11.0)
WX-251	22	46 3⁄4	62.0	0.55	22.9	19.2	16.7	92 (13.9)
WX-302	26	47 3⁄16	86.0	0.54	31.8	26.7	23.2	123 (18.9)
WX-350	26	61 7⁄8	119.0	0.39	44.0	36.9	32.1	166 (24.5)
WX-302-HC	26	47 3⁄16	86.0	0.54	—	—	—	125 (18.9)
WX-350-HC	26	61 7⁄8	119.0	0.39	—	—	—	168 (24.5)

Precharge Pressure for WX-104-S thru WX-203 is 30 PSIG and Sys. Conn. is 1" NPTF.
Precharge Pressure for WX-205 thru WX-350-HC is 38 PSIG and Sys. Conn. is 1 1⁄4" NPTF.
Maximum Working Pressure is 100 PSIG and Maximum Working Temperature is 200° F.

Note: Drawdown can be affected by various ambient and system conditions, including temperature and pressure.

8-32 *Stand-type pressure tank.* Amtrol, Inc.

In-line tanks

In-line tanks (Fig. 8-33) are designed to be installed right off of a water pipe. They might be suspended below the pipe, or they might rise above the pipe. The size of in-line tanks vary. A brand that I use offers in-line tanks with capacities of 2 gallons, about 4½ gallons, about 8½ gallons, a little over 10 gallons, and 14 gallons. This range of sizes can be adequate for small homes.

An in-line tank does not normally mount on a bracket or sit on a floor. It typically either hangs from the pipe or sticks up above the pipe. When floor space is at a premium, an in-line tank is desirable. As a builder, I usually try to keep the minimum size up in the ten-gallon range. However, I recently installed one of the 4½-gallon models in a summer cottage. The cottage had only one bathroom and rarely accommodated more than two people, so the small tank was deemed adequate. You must match your tank to your needs.

Stand models

Stand models (Fig. 8-34) are the type of pressure tank most often installed in homes. These units sit on a floor in a free-standing mode. Piping is run from the pump to the tank and then from the tank to the water distribution system. With capacities ranging from about 10 gallons up to 119 gallons, this type of tank can meet any residential need.

I have a stand model in my home. It's capacity is around 40 gallons. This tank costs more than a smaller tank, but its size allows my pump to run less often. Since it runs less, it saves electricity and prolongs the life of my pump. I think the trade-off is worthwhile.

A stand model is easiest to install when a tank tee (sometimes called a tank cross) is used (Fig. 8-35). This fitting is designed for use with a pressure tank. The tank tee screws into the pressure tank and provides both an inlet and outlet opening. In addition to these openings, the fitting is tapped for accessories, such as pressure gauges and relief valves. The tank tee not only makes the job easier, it gives the appearance of a more professional installation (Fig. 8-36).

Pump-stand models

Pressure tanks are available in pump-stand models (Fig. 8-37). These tanks are frequently used in conjunction with jet pumps. They are not needed with submersible pumps, since submersibles are hung in a well. A pump-stand model is designed to sit horizontally. It has a bracket on top and the pump is bolted to the bracket.

7 bar series

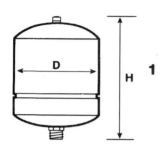

Model	Dimensions		Total
No.	Diameter	Height	Volume
	mm	mm	Ltr
WX 2,6	156	228	2,6
WX 4	156	302	4,1
WX 8	200	320	8
WX 18	280	380	18
WX 33	280	630	33

8-33 *Specifications for in-line pressure tanks.* Amtrol, Inc.

Pump-stand models offer a way to conserve floor space. Since a pump attaches to the top of the stand, there is only one component sitting on the floor. With a typical stand model, both the pressure tank and the pump are installed on the floor. The manufacturer I deal with has two sizes of pump-stand tanks. One is about 8½ gallons, the other is 14 gallons. I would opt for the latter.

Underground tanks

Although I have never installed one, underground tanks are available (Fig. 8-38). The manufacturer I use offers them in sizes ranging from 14 to 62 gallons. I can't recall a situation when an underground tank would have helped me, but I'm sure there must be occasions when they are desirable.

Diaphragm tanks

Diaphragm pressure tanks are common in today's plumbing systems (Fig. 8-39). This was not always the case. Older houses often have plain old galvanized storage tanks. These standard tanks can be a real pain. They frequently become waterlogged. By this, I mean that they lose their air content and fill with water. When this happens, the tank must be drained, pumped up with air, and refilled with water. A waterlogged tank makes a pump run every time water is called for at a faucet, thus eliminating the advantage of having a pressure tank.

Modern pressure tanks come precharged with air, and they have a diaphragm system that eliminates waterlogging. Standard tanks are still available, but they are rarely used. I suggest that you specify a diaphragm tank in your well specs.

Rust is another problem with a standard tank. After some period of time, a metal holding tank begins to rust. Air leaks out and then,

1,5/3,0 bar	2,0/3,5 bar	2,5/4,0 bar	System Connection	Precharge Pressure	Shipping Wt./Vol.
	Drawdown				
Liter	Liter	Liter	R"	bar	KG / m³
1,0	0,9	0,8	3/4	1,5	1,0 / ,005
1,5	1,4	1,2	3/4	1,5	1,5 / ,007
3,0	2,6	2,4	3/4	1,5	2,3 / ,02
6,7	6,0	5,4	3/4	1,5	4,1 / ,03
12,4	10,9	9,9	3/4	1,5	6,8 / ,05

8-33 *Continued.*

eventually, water leaks. This means a patch or replacement must be made. Modern pressure tanks have liners so that water never comes into contact with the metal housing. The liner eliminates the possibility of interior rust. Bacteria growth and rust in drinking water are not nearly as likely with a lined tank as they are with a traditional metal tank.

Sizing

Determining the correct size for a pressure tank is an important step in designing and installing a good pump system. The pressure tank protects the pump. I mentioned earlier that you must consider the number of people and plumbing fixtures when sizing your pressure tank. My comment was meant to get you thinking about how little water use it takes to drain down a small tank.

If you want to size a tank, you should work from the specifications for the pump that is to be installed. In other words, the ideal sizing comes from carefully matching your tank to your pump. The number of people or plumbing fixtures plays only a small role, if any, in determining tank size.

Pressure tanks are designed to work with pressure switches. These switches can be set for various cut-in and cut-out pressures. It has long been common for a pump to cut on when tank pressure drops to 20 pounds per square inch (psi). At this cut-in rate, a typical cut-out rate is 40 psi. For years, this has been something of a standard in well systems. However, times change, and so do standards.

It is not at all uncommon for cut-in pressures to be set at 30 psi today. A corresponding cut-out pressure is 50 psi. Some homes have well systems set up with cut-in pressures of 40 psi and cut-out pres-

10 bar series

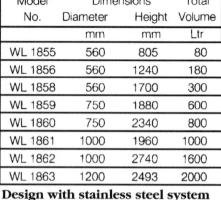

| Model | Dimensions | | Total |
No.	Diameter	Height	Volume
	mm	mm	Ltr
WL 1855	560	805	80
WL 1856	560	1240	180
WL 1858	560	1700	300
WL 1859	750	1880	600
WL 1860	750	2340	800
WL 1861	1000	1960	1000
WL 1862	1000	2740	1600
WL 1863	1200	2493	2000

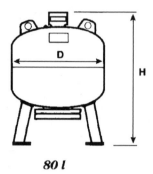

80 l

Design with stainless steel system connection (V4A)

WL 1935	560	805	80
WL 1936	560	1240	180
WL 1938	560	1700	300
WL 1939	750	1880	600
WL 1940	750	2340	800
WL 1941	1000	1960	1000
WL 1942	1000	2740	1600
WL 1943	1200	2493	2000

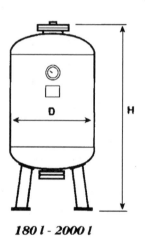

180 l - 2000 l

16 bar series

WL 1955	560	805	80
WL 1956	560	1240	180
WL 1958	560	1700	300
WL 1959	750	1880	600
WL 1960	750	2340	800
WL 1961	1000	1960	1000
WL 1962	1000	2740	1600
WL 1963	1200	2493	2000

Maximum Operating Temperature = 90°C.
Horizontal designs and tanks for Operating Pressures of 25 bar are available on request.
Note: Drawdown can be affected by ambient and system conditions, including temperature and pressure.

8-34 *Specifications for pressure tanks with replaceable bladder designs.*
Amtrol, Inc.

| 1,5/3,0 bar | 2,0/3,5 bar | 2,5/4,0 bar | System Connection | Precharge Pressure | Shipping Wt./Vol. |
| Drawdown | | | | | |
Liter	Liter	Liter	R"	bar	KG / m3
30	27	24	2	3,5	59 / ,25
68	60	54	2	3,5	83 / ,39
113	99	90	2	3,5	155 / ,53
225	198	180	2	3,5	285 / 1,06
300	264	240	2	3,5	360 / 1,32
375	330	300	3	3,5	400 / 1,96
600	528	480	3	3,5	540 / 2,74
750	660	600	3	3,5	780 / 3,59

30	27	24	2	3,5	59 / ,25
68	60	54	2	3,5	83 / ,39
113	99	90	2	3,5	155 / ,53
225	198	180	2	3,5	285 / 1,06
300	264	240	2	3,5	360 / 1,32
375	330	300	3	3,5	400 / 1,96
600	528	480	3	3,5	540 2,74
750	660	600	3	3,5	7890 / 3,59

30	27	24	2	3,5	64 / ,25
68	60	54	2	3,5	102 / ,39
113	99	90	2	3,5	220 / ,53
225	198	180	2	3,5	400 / 1,06
300	264	240	2	3,5	505 / 1,32
375	330	300	3	3,5	560 / 1,96
600	528	480	3	3,5	756 / 2,74
750	660	600	3	3,5	1330 / 3,5

8-34 *Continued.*

8-35
Tank tee. Goulds Pumps, Inc.

AV20-10 (1") tank cross center
AV20-12 (1^1/$_4$") tank cross center:
Designed for use with aqua air tanks.

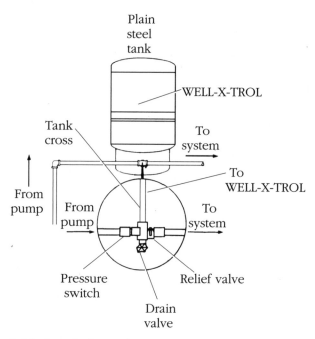

8-36 *Detail of a tank-tee set-up.* Amtrol, Inc.

Pump Stand models

Model No.	Dimensions		Total Volume (gals)	Max. Accept. Factor	Drawdown			Shipping
	Diameter (ins)	Height (ins)			20/40 (gals)	30/50 (gals)	40/60 (gals)	Wt. (Vol.) lbs (cu ft)
WX-103-PS	12 1/2	11 1/4	8.6	0.28	3.1	2.7	2.2	20 (2.4)
WX-200-PS	17 1/4	15 5/8	14.0	0.81	5.2	4.3	3.8	29 (4.0)

Precharge Pressure is 30 PSIG and Sys. Conn. is 3/4" NPTM Fitting for 103-PS and 1" NPTF Coupling for 200-PS.
Maximum Working Pressure is 100 PSIG and Maximum Working Temperature is 200° F.

8-37 *Pump-stand type pressure tank.* Amtrol, Inc.

Underground models

Model No.	Dimensions Diameter (ins)	Height (ins)	Total Volume (gals)	Max. Accept. Factor	Drawdown 20/40 (gals)	30/50 (gals)	40/60 (gals)	Shipping Wt. (Vol.) lbs (cu ft)
WX-200-UG	15 3⁄8	22	14.0	0.81	5.2	4.3	3.8	25 (3.8)
WX-202-UG	15 3⁄8	29 3⁄4	20.0	0.57	7.4	6.2	5.4	33 (4.9)
WX-250-UG	22	33 3⁄8	44.0	0.77	16.3	13.6	11.9	63 (11.0)
WX-251-UG	22	44 1⁄2	62.0	0.55	22.9	19.2	16.7	83 (13.9)

Precharge Pressure for Models 200-UG and 202-UG is 30 PSIG and Sys. Conn. is 1" NPTF Coupling.
Precharge Pressure for Models 250-UG and 251-UG is 38 PSIG; Sys. Conn. is 1 1⁄4" NPTF Coupling.
Maximum Working Pressure is 100 PSIG and Maximum Working Temperature is 200° F.

8-38 *Underground pressure tank specifications.* Amtrol, Inc.

1.

WELL-X-TROL has a sealed-in air chamber that is pre-pressurized before it leaves our factory. Air and water do not mix eliminating any chance of "waterlogging" through loss of air to system water.

2.

When the pump starts, water enters the WELL-X-TROL as system pressure passes the minimum pressure precharge. Only usable water is stored.

3.

When the pressure in the chamber reaches maximum system pressure, the pump stops. The WELL-X-TROL is filled to maximum capacity.

4.

When water is demanded, pressure in the air chamber forces water into the system. Since WELL-X-TROL does not waterlog and consistently delivers the maximum usable water, minimum pump starts are assured.

8-39 *How diaphragm pressure tanks work.* Amtrol, Inc.

sures of 60 psi. The increase in pressure is due to many factors. People often like to have increased water pressure at some plumbing fixtures, such as showers. Also, some modern plumbing fixtures and devices require higher working pressures than they did in the past. This all contributes to the trend for higher pressures.

Before you can size a pressure tank, you must determine your cut-in and cut-out pressures. You must know the gpm rating of the pump. Also, you need to determine how many start-ups the pump can be subjected to in a 24-hour period (Fig. 8-40). In addition, the minimum run time for the pump must also be determined. The minimum run time is usually established by the manufacturer, so check your pump paperwork.

Once you know all the variable mentioned above, you can select a tank of the proper size. However, you need some type of sizing table to do this math. Most manufacturers of pressure tanks are happy to provide sizing tables. As an example, I'm providing you with a sizing table (Fig. 8-41) and a sample worksheet (Fig. 8-42). You can see for yourself how easy it is to make sure that the homes you build are provided with suitable pressure tanks.

Installation procedures

Installation procedures for pressure tanks can take many forms. You might install an underground tank to work in conjunction with a sub-

Table 1 number of starts

Motor rating	Maximum starts per 24 hr. day	
	Single phase	Three phase
Up to 3/4 hp	300	300
1 hp thru 5 hp	100	300
7 1/2 hp thru 30 hp	50	100
40 hp and over		100

8-40 *Recommended maximum number of times a pump should start in a 24-hour period.* Amtrol, Inc.

Sizing table

Pump discharge rate gpm (approx.)	Operating pressure—psig					
	20/40		30/50		40/60	
	ESP I	ESP II	ESP I	ESP II	ESP I	ESP II
2.5	WX-104	WX-201	WX-104	WX-202	WX-104	WX-202
5	WX-201	WX-205	WX-202	WX-205	WX-202	WX-250
7	WX-202	WX-250	WX-203	WX-251	WX-205	WX-251
10	WX-203	WX-251	WX-205	WX-302	WX-250	WX-302
12	WX-205	WX-302	WX-250	WX-302	WX-251	WX-350
15	WX-250	WX-302	WX-251	WX-350	WX-251	WX-350
20	WX-251	WX-350	WX-302	(2)WX-251	WX-302	(2)WX-302
25	WX-302	(2)WX-302	WX-302	(2)WX-302	WX-350	(3)WX-251
30	WX-302	(2)WX-302	WX-350	(1)WX-302 (1)WX-350	WX-350	(2)WX-350
35	WX-350	(1)WX-302 (1)WX-350	WX-350	(2)WX-350	(2)WX-251	(3)WX-302
40	WX-350	(2)WX-350	(2)WX-251	(3)WX-302	(2)WX-302	(1)WX-302 (2)WX-350

8-41 *Sizing and selection information for pressure tanks.* Amtrol, Inc.

mersible pump. A pump-stand tank is sometimes a good choice for a jet pump. In-line tanks can also come in handy. Stand models are the type used most often (Fig. 8-43). A multiple-tank set-up (Fig. 8-44) might be called for, although it is highly unlikely in a residential system unless you are building a house on a farm. Your installer should know how to make a good installation, and the diagrams I've provided should help you understand the procedures.

The relief valve

Make sure a relief valve is installed to protect the pressure tank from excessive pressure. A disaster could result if this fitting is omitted. A relief valve is a code requirement, so someone should be checking to make sure one is installed. But don't take this chance, always take the time to check the installation yourself.

Ratings on pressure tanks and relief valves can vary, but most tanks are rated at 100 psi and the relief valves used with them are rated at 75 psi. If a relief valve isn't installed, a pressure tank can blow up, causing personal and property damage. The same thing can happen if the relief valve is rated higher than the tank's working pressure.

WELL-X-TROL QUICK SIZING FORM
(We suggest you make an office copy of this page when ready to calculate.)

For selecting WELL-X-TROLs for a different running time than ESP I or ESP II, and/or at pressure ranges the same or different than 20/40, 30/50, 40/60:

THINGS YOU MUST KNOW

1. System flow rate (pump capacity or discharge) _____ GPM

2. Desired running time, in minutes and fractions of minutes (1.5 min. = 1 min. 30 sec.) _____ Min.

3. Pump cut-in, in gauge pressure _____ Psig

4. Pump cut-out, in gauge pressure _____ Psig

CALCULATING TANK SIZE

5. Multiply Line 1 by Line 2 and enter ESP Volume _____ ESP Vol.

6. Refer to Table 1. Find pressure factor for Line 3 and Line 4 and enter _____ P.F.

7. Divide Line 5 by Line 6 and enter minimum total WELL-X-TROL Volume _____ Gals.

8. Refer to Table 2 and select WELL-X-TROL model that is greater than Line 7 for "Total Volume" and Line 5 is less than "Maximum ESP Volume" _____ WX No.

9. Select precharge pressure _____ Psig

> NOTE: The precharge pressure *must* be adjusted to the pump cut-in pressure. For the following example reduce from 40 to 25 psig.

> EXAMPLE: A system flow will be delivered by a pump at a rate of 12.5 GPM. The pump switch is to be installed at the WELL-X-TROLL and has been determined to cut-in the pump at 25 psig. Its differential, or operating range, is 20 psi. It is desired to have the pump run *at least* one minute and 30 seconds every time it starts. Which WELL-X-TROL will provide "ESP"?

THINGS YOU MUST KNOW

1. System flow rate (pump delivery) 12.5 GPM

2. Desired running time, in minutes and fractions of minutes (1.5 min. = 1 min. 30 sec.) 1.5 Min.

3. Pump cut-in, in gauge pressure 25 Psig

4. Pump cut-out, in gauge pressure 45 Psig

CALCULATING TANK SIZE

5. Multiply Line 1 by Line 2 and enter ESP Volume 18.8 ESP Vol.

6. Find Pressure factor for Line 3 and Line 4 in Table 1, and enter 34 P.F.

7. Divide Line 5 by Line 6 and enter minimum total WELL-X-TROL volume 55.2 Gals.

8. Refer to Table 2 and select WELL-X-TROL model that is greater than Line 7 for "Total Volume" and Line 5 is less than "Maximum ESP Volume" WX-251

A WX-251 has a total volume of 62 gallons and a maximum ESP volume of 34 gallons.

8-42 *Pressure tank sizing form.* Amtrol, Inc.

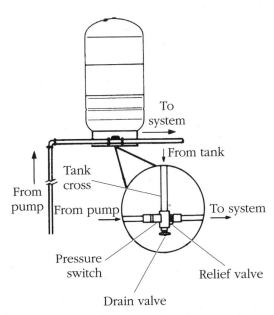

To
system

From tank

Tank
cross

From
pump

From pump

To system

Pressure
switch

Relief valve

Drain valve

8-43 *Tank tee being used with a stand-type pressure tank.* Amtrol, Inc.

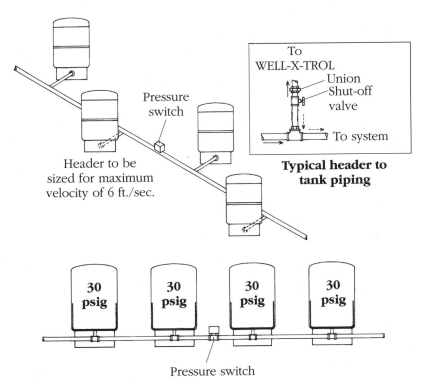

Pressure
switch

Header to be
sized for maximum
velocity of 6 ft./sec.

To
WELL-X-TROL
Union
Shut-off
valve

To system

**Typical header to
tank piping**

| 30 psig | 30 psig | 30 psig | 30 psig |

Pressure switch

8-44 *Diagram of multiple pressure tanks being installed together.*
Amtrol, Inc.

This is a serious issue, so take an active part in making sure the right relief valve is installed properly.

Keep it dry

A pressure tank lasts longer when you keep the exterior dry. In some cases, this might require installing a tank on blocks or a platform. It's a good idea to elevate the tank if it's installed in an area that tends to get wet, such as a basement or cellar. This is very simple to do at the time of installation, but it becomes a bit complicated after the fact. Spending a few bucks for some blocks can save money for your customers down the road. Happy customers are what successful businesses are all about, so don't forget them when you are putting out the specs on a job.

Are you confused yet?

Are you confused yet? Do you know what a shallow well is? Can you tell the difference between a dug well and a drilled well? Is a bored well the same as a dug well? Hey, don't take these questions too seriously. I suspect that you now know more about shallow wells than many plumbers.

Our next chapter deals with alternative water sources. We are going to talk about springs, ponds, lakes, and so forth. Also, we are going to discuss alternative well and pumping methods. If you're ready, let's get into Chapter 9.

9

Alternative
water sources

Alternative water sources, for our purposes, are any water source that is not a dug, bored, or drilled well. Municipal water sources are excluded from the alternative category. Springs, lakes, cisterns, driven wells, and similar sources of water are the types that I call alternatives. In some cases, the water does not have to be safe for drinking.

For example, let's say you are building an exclusive home for an owner who enjoys horses. The homeowner might want water available for the stables, and this water doesn't necessarily have to be safe for human consumption. If the stable chores require a lot of water, it could put a strain on the residential well. Rather than deplete the home's potable drinking water, it might be practical to put the horse and stable water on a separate system. You might even suggest a solar-powered system (Figs. 9-1 and 9-2).

Even if you don't have a need for nonpotable water, an alternative water source might be worth considering. Many homes get their water from springs and driven wells. While these two sources of water are not normally considered to be dependable for a high quantity of water, a lot of houses manage to get by with them.

Irrigation systems for lawns and gardens can use nonpotable water. Like the horse example, irrigation is a prime candidate for a separate water source. The amount of water used for irrigation can be substantial, and it is usually during dry seasons that water is most needed. Does it make sense to avoid using potable well water during a dry spell? Of course it does, but an alternative water source can be a solution. There are, in fact, many reasons why a second water source might be wanted.

An alternative water system can produce potable water. Springs are an excellent example. Have you noticed all the bottled spring water that is sold in your local grocery store? Many people feel that the

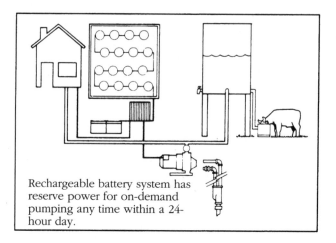

9-1 *Solar-powered pump system with a rechargeable battery system.* A.Y. McDonald MFG. Co., Dubuque, Iowa

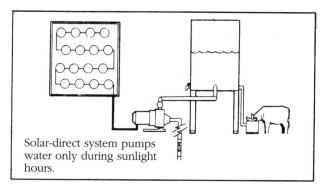

9-2 *Solar-powered pump system without a rechargeable battery system.* A.Y. McDonald MFG. Co., Dubuque, Iowa

only good drinking water is spring water. If it's good enough to serve in fancy restaurants, it's good enough to drink at home. However, some special precautions should be exercised to avoid contamination of a spring. We are going to talk more about this later in the chapter.

Driven wells can produce enough water to serve a full-time residence. Due to their design, driven wells can't be considered a dependable source of water when a lot is needed, but they are very inexpensive to install. They have limitations, but you might find that a driven well is just what you need.

Have you ever worked on an old home that had a large cistern installed in the cellar? They are quite common in many of the older homes in Maine. I never saw one in Virginia, but I've seen several during my plumbing and remodeling work in Maine. Cisterns were used a lot in the old days and they can still be useful today.

As we move through this chapter, you are going to learn a lot about various types of water sources. Without a doubt, conventional wells are going to be your most likely source of water when building a house in the country. But, alternative water sources have their place, and this is the place to learn about them.

Driven wells

Driven wells are very inexpensive to install. If ground conditions are suitable, a driven well is easy to make. No fancy truck or drilling rig is needed. A ladder, a sledge hammer, a few specialized parts, along with a strong back and arms, are all that are needed to make a driven well.

What is a driven well? It is a pipe driven into the ground. Some people refer to these wells as driven points or just as points. Many of these wells exist in parts of Maine. They are especially popular along the coast, where summer cottages are built on sandy soil.

Personally, I wouldn't choose a driven well to supply a house with water when the property is to be used as a full-time residence. I do, however, know of homes where a point provides the only source of potable water for entire families. Driven points have a very small reserve capability. If the water source for the well is strong, it can produce a lot of water. But, if the water source is slow, it's easy to run this type of well dry. Let me give you some specifics on driven wells.

The components of a driven well are simple. The first piece is the drive point, which also serves as a filter (Fig. 9-3). Pipe used to form the well makes up the second component, and the last piece of equipment is a pump. The well pipe usually doesn't exceed 2 inches in diameter, and 1½-inch-diameter pipes are common. Under ideal conditions, this type of well can be driven to a depth of about 50 feet.

The point

The point you select might be made of reinforced steel or it could have a bronze tip. The pointed end allows it to be driven into soft ground. At the opposite end of the point are pipe threads that permit additional sections of pipe to be added as the point is driven. These threads are protected by a drive cap during the driving phase of the work.

Well points are available with different types of screens. The screens act as a filter. The ground conditions dictate the type of screen

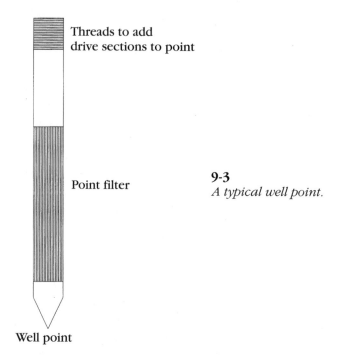

Threads to add
drive sections to point

Point filter

9-3
A typical well point.

Well point

to use for a particular well. For example, you would use a screen with a wide mesh if you were pulling water from coarse gravel. A fine-mesh screen would be used if water was being obtained from sand. The openings in the filter must be matched to the ground conditions to avoid having sand and similar particles enter the well.

The well pipe

The well pipe you choose to use might be standard, galvanized water pipe or special piping designed just for driven wells. Go with the specialized piping. Galvanized pipe tends to rust along the interior sides of the pipe, as well as at the threads. Since threads have thin walls, it's not unusual for this area to be the first point of deterioration. I don't recommend galvanized pipe for several reasons, and rust is one of them.

Another problem with galvanized pipe is that it doesn't stand up well to heavy pounding. To get the pipe short enough to drive effectively, it must be cut and then threaded. You can certainly make your own drive sections in this way, but the work is time consuming and hard on the arms, unless you have a power threader. Since galvanized pipe does not drive well, the well depth is often limited to about one-half the depth of a well constructed with specially-designed drive pipe.

To drive a pipe with power, it must not be too long. A 5-foot section of pipe usually works well. A platform is needed for the first few whacks with the hammer, but then a person can move down to the ground as the pipe is driven in deeper. The special drive pipe that I'm talking about is often called a riser section. Riser sections run about 5 feet in length, which makes them ideal for their purpose.

Drive caps

Drive caps protect the pipe threads during the driving process. If you don't use a drive cap, the end of the pipe is going to roll out as it is hit, making it impossible to thread couplings onto the pipe. Drive caps are available with either male or female threads. The caps are quite simple, but very necessary.

Power driving

Power driving can produce a well depth of about 50 feet. Using a sledge hammer, which is the standard driving technique, probably nets only about 30 feet of depth, although that can sometimes be more than enough. A number of set-ups exist for using mechanical driving power. Impact drivers are also available, and they can make driving a well easier. If you were to go into business as a professional installer, you would probably want to have a mechanical means to power drive wells. However, the majority of people that I know drive wells the old-fashioned way, with a sledge hammer and human sweat.

Getting it out

Getting a drive pipe into the ground can be difficult, but getting it out can be even more troublesome. When you drive a well point, you never know when some buried obstruction is going to stop your best efforts. When this happens, you must remove the riser sections and point to start somewhere else. You won't have much success doing this by hand if the point has reached a significant depth.

Pipe clamps are one solution to the problem of pipe removal. Putting the clamps around the pipe can allow you to drive the pipe out of the hole with your hammer. Lifting jacks (such as the type used to jack up automobiles) can also be used, in conjunction with pipe clamps, to raise sections of pipe. Blocks of wood should be placed on the ground to keep the jacks from sinking. The jacks exert upward pressure on the pipe clamp, thereby lifting the pipe. You must frequently reset the jacks and clamps, but this method certainly can pull the pipe out of the ground. Other methods of removal, such as the use of winches, can also be employed to remove pipe.

The actual driving

The actual driving of a well point should be started slowly. Many people use posthole diggers or an auger to get a hole started, up to a depth of about two feet. Once the hole is dug, the well point is set in the hole. I should add that some type of pipe sealant compound must be applied to the male threads of all joints to make them watertight. It's essential for the well point and first section of well pipe to be driven into the ground as close to perfectly vertical as is possible. A carpenter's level can be used to check for the vertical position. You should only tap the drive cap during this stage of the operation—don't bang it. Use light strokes so that you can control the position of the pipe.

As the first section gets into the ground, you must remove the drive cap, add pipe dope sealant to the male threads on the riser, then add a new riser to the rig. Install the drive cap to the end of the new riser and resume driving. As the first section disappears from sight, you can increase the power of your pounding. However, don't hit the pipe if it's vibrating. This could cause damage to underground sections.

The sequence of events for driving a well point are routine. You drive a section to a point near the top of the ground. Then you remove the drive cap, dope the threads and add a new section. Next, replace the drive cap and keep driving. This procedure continues until you can go no further or until you have an abundant water supply.

Terminating the well

You must decide how the well is going to terminate. Will you install a pitless adapter? This is a good idea for a well that has its water service running below the frost line. Will you install a simple pitcher pump on top of the pipe? This can be a suitable solution for seasonal use. Your well can be terminated above or below grade. Elbow fittings can allow you to turn the well pipe horizontally, if you desire. Plan your termination in advance.

Suitable soils

Suitable soils must exist in order to drive a well. Not all ground conditions allow a well point to be driven, even with mechanical assistance.

It is not possible, for example, to drive a point through bedrock. Soft, moist clay is a good soil for driving a well. Coarse sand and gravel are also good. You can drive a point through fine sand and hard clay, but the work is difficult. If you can do some test borings before sinking a point, it can certainly prove helpful.

Quantity

There is no way to predict the quantity of water that can be produced by a well point. It could be 50 gallons of water, or 250 gallons. The recovery rate is equally impossible to predict. Until you sink the point and test the well, there is no way to know. However, the combination of a driven well and a large pressure tank can produce adequate water for a full-time residence that is occupied by a family.

Quality

The quality of water produced from a driven well can be compared to that of water coming from a dug or bored well. Since these types of wells normally connect with water at about the same depths, the type of water that's produced should be about the same. All water that's intended for human consumption should be periodically tested to make sure its safe to drink.

Cisterns

Cisterns can be considered holding vessels for water. They might be constructed above ground or below grade. Brick, stone, or concrete can be used to build a cistern. Such a water holding area can be small enough to water a family vegetable garden or large enough to provide water for a family during several months of dry times.

Cisterns typically are filled by surface water. For this reason, the water is not normally considered acceptable for drinking. However, if filtering and treatment systems are used in conjunction with a cistern, the quality of the water can be raised to a potable level. Additionally, a cistern can be filled from a potable water source, such as a well, to create a large reserve of drinkable water. How long this water can be kept in a potable condition varies with conditions, but this is one way to make the most of a well during times of plentiful water.

In older homes, it was not uncommon for large cisterns to be built beneath the first floor of the structures. Some of these collection tanks resemble modern-day, above-ground swimming pools in size. In fact, a modern, above-ground pool can serve as a very effective cistern. Let me give you a quick example before it slips my mind.

I purchased an above-ground pool so that my young daughter could learn to swim. The pool is about 42 inches deep, and it has a diameter of 15 feet. According to the paperwork provided with the pool, its water capacity is 4400 gallons. This pool cost less than $300. Setting the pool up took only a few hours. If this holding tank was being used as a cistern, it would be quite effective, very efficient, and low in cost. Depending upon your cistern needs, don't overlook the possibility of above-ground pools as a material choice.

Many of the cisterns I've seen have been built in the cellars of homes. Some of these collection vessels were made to have water pumped into them. Many of them collected rainwater from the roof of the house. A lot of water runs off the roof of a house during a rainstorm. If this water is diverted to a cistern, it doesn't take long to build up several hundred gallons of reserve water. It's entirely possible for a cistern to collect and contain enough water to provide a family with water for up to six months. In areas where rainfall comes in spurts, a cistern can be a lifesaver, so to speak.

Water from a cistern can be pumped in a manner similar to that used with shallow wells. This water can serve domestic needs when it's potable. Irrigation and farm animals are two reasons for using a cistern. Roof water can be easily collected using gutters and piping. A cistern might be filled from a stream during wet times so that water would be available in dry times, when the stream has dried up.

Cisterns are rarely considered a potable water source for a new house. This type of water collection is not comparable to a drilled well. However, there are many circumstances that might warrant the use of a cistern.

Ponds and lakes

The use of existing ponds and lakes can do a lot for a property when high volumes of water are needed. One such situation might be a home where the owner operates a commercial greenhouse. A lot of water is needed to grow plants commercially. If a lake or pond is handy, the expanse of reserve water can be used to keep the plants healthy. Such a water source can also be used for irrigation or watering livestock.

Ponds and lakes are not normally suitable as a source of potable water, unless the water is purified. This is certainly possible, but the expense and trouble often outweighs the cost of drilling a well. For the most part, ponds, lakes, streams, and rivers should be ignored as potable water sources.

Springs

Springs have long been a source of drinking water. However, the water is not always safe to drink. If a spring is to be used as a primary source of potable water, some special provisions should be made to ensure the quality. For instance, the spring should be protected from surface water.

If a spring is to be used for potable water, a watertight container of some type should be placed around the spring. Well casing works fine. Springs located on the sides of hills should have diversion ditches installed above them. Ditches help keep surface water from running into the spring. Slotted pipe and gravel can be installed in trenches to intercept and divert surface water.

Fencing should be installed around a spring to prevent the entry of any animals. The uphill side of the spring should be fenced for a distance that is sufficient to prevent animal activity from causing contamination. Even with these protections, a spring should be tested regularly to assure its potability.

Springs, when available, are an inexpensive water source. They can provide large quantities of water, and the quality can be quite good. Since springs are not abundant, they cannot be counted on as a water source. But, if one is available, it can provide numerous benefits to the property owner.

I remember visiting a spring on nearly a daily basis during the summers of my childhood. Going down to the spring was not a necessity. My parents and grandparents had municipal water supplies in their homes. I used to go to the spring as a pleasure. The spring was used as a water source by some people in the neighborhood. Looking back on this spring, I remember it having a metal lining. It might have been a 55-gallon drum. A drum probably wouldn't be considered a good lining, but as I recall, that's what was used. To the best of my knowledge, no one ever got sick drinking from this well. This doesn't mean that I would recommend that you use or install such a water source.

Here in Maine, I live near a spring that is visited daily by dozens of people. At times, people are lined up to get access to the spring water. Some of these people have a great number of half-gallon, plastic jugs with them to collect the spring water. Unlike the spring I used as a child, this one has a pipe that extends out of the side of a hill. Water runs from it constantly. I've never tried the water, but a great number of people do. As often as I drive by this spring, I can recall only one or two occasions when the water wasn't running.

A gentleman I know gets all of his domestic water from a spring. He has done this for a long time. Last winter his water supply pipe froze. The spring didn't, but his water service did. When you consider that a spring in Maine can handle the winter temperatures without freezing, it's certainly possible for a spring to provide year-round water to a home. Still, you can't count on a spring having enough constant flow to be a primary water source.

Many needs

Alternative water sources can fill many needs. They can provide water for all domestic uses. Livestock can be watered with alternative sources. Gardens and lawns can be watered with nonpotable water. Fire protection can be boosted if a pond or cistern is available. Water-source heat pumps can be fed by alternative water sources and wells. Because so many potential uses for water exist, alternative water sources should always be considered.

As a builder, you should be aware of the many options available to you when choosing water systems for the homes you build. Conventional wells are going to continue to provide most of the water that's used in rural locations, but other natural resources and man-made provisions can enhance the living conditions of people in the country. With a little research and some creative thinking, you can offer customers some water options that other builders might not consider. This effort can help you win more bids and make more money.

10

Gravel-and-pipe septic systems

When discussing septic systems, gravel-and-pipe systems are about the least expensive option available. Just because these systems are less expensive than other types of systems doesn't mean that they are of a lower quality or don't perform as well. The reason pipe-and-gravel systems cost less is simple, they don't require as much material or time to build. Why is this? It's because they are installed in better quality soil. Soil of poorer quality requires a more complex system, such as a chamber system.

If you build houses on spec, you should be particularly alert to the type of septic system your houses requires. When people buy a new house, they don't normally care whether the septic system is of a pipe-and-gravel type or a chamber type, as long as the system meets code requirements and is designed and installed to give years of worry-free service. If you have two houses sitting side by side, one with a chamber system and one with a pipe-and-gravel system, no one knows the difference unless you tell them. However, potential buyers are going to notice one big difference, and that's the price. A chamber system can easily cost twice that of a pipe-and-gravel system. To make the same profit, the house with the chamber system must be priced thousands of dollars higher than the other house.

When I was a builder in Virginia, I never had to install a chamber system. All of the septic systems installed through my building company were pipe-and-gravel systems. Now, as a builder in Maine, I see a lot more chamber systems than I do pipe-and-gravel systems.

How much difference is there in the cost between a pipe-and-gravel system and a chamber system? The differences vary. I'm sure costs in Maine are not comparable to costs in Nebraska or California. From my personal experience, an average pipe-and-gravel system in Maine costs, at builder's rates, a little over $4000. A chamber system for the same type of house comes in somewhere around $8500, although I've seen them at more than $10,000. This is a huge difference in cost. And, the only reason for this increase is the suitability of the soil where the system is installed.

Maine has a lot of wet land and a lot of bedrock. Both of these are factors in forcing the use of chamber systems. I was lucky when I built my house because I was able to use a pipe-and-gravel system. Was this all luck? Not exactly. Knowing that a chamber system would cost at least twice as much as a simple septic system, I considered this fact in my search for land. You can use this same approach when you are looking for lots and land on which to develop spec houses.

Consider what I'm about to tell you. I was involved in a fairly large land development project some years back. The building lots were in a rural area and consisted of 20 acre building sites. Each lot was required to have its own septic system. The land perked well, and all the lots were approved for pipe-and-gravel systems. With this particular project, we were dealing with ten lots. If those lots had required more expensive septic systems, we might have lost $40,000 or more in sales potential. So you see, the difference between a pipe-and-gravel system and a complex system can have a lot of impact on your profit picture.

The components

Let's talk about the basic components of a pipe-and-gravel septic system (Fig. 10-1). Starting near the foundation of a building, there is a sewer. The sewer pipe should be made of solid pipe, not perforated pipe. I know this seems obvious, but I did find a house about three years ago that was installed with perforated drain-field pipe. It was quite a mess. Most jobs today use schedule-40 plastic pipe for the sewer. Cast-iron pipe can be used, but plastic is the most common and is certainly acceptable.

The sewer pipe runs to the septic tank. Septic tanks can be made of many types of materials, but most of them are concrete. It is possible to build a septic tank on site, but every contractor I've known uses precast tanks. An average size tank holds about 1000 gallons. The connection between the sewer and the septic tank should be watertight.

The discharge pipe from the septic tank should be made of solid pipe, just like the sewer pipe. This pipe runs from the septic tank to a distribution box, which is also normally made of concrete.

The drain field (also called a leach field) is constructed according to an approved septic design. In basic terms, the excavated area for

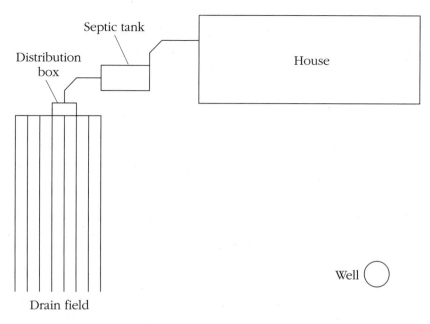

10-1 *Typical site plan.*

the septic bed is lined with crushed stone. Perforated plastic pipe is installed in rows. The number of drain pipes and the distance between them is controlled by the septic design. All of the drain-field pipes connect to the distribution box. The septic field is then covered with material specified in the septic design.

As you can see, the list of materials is not a long one. Some schedule-40 plastic pipe, a septic tank, a distribution box, some crushed stone, and some perforated plastic pipe are the main ingredients. This is the primary reason why the cost of a pipe-and-gravel system is so low when compared to other types of systems.

Types of septic tanks

Many types of septic tanks are in use today. Precast concrete tanks are, by far, the most common. However, they are not the only type of septic tank available. For this reason, let's discuss some of the material options that are available.

Precast concrete

Precast concrete, as I've already said, is the most popular type of septic tank. When this type of tank is installed properly and not abused, it can last almost indefinitely. However, heavy vehicular traffic running over the tank can damage it, so this should be avoided.

Metal

Metal septic tanks were once prolific. A great number of them are still in use, but new installations rarely involve a metal tank. The reason is simple, metal tends to rust, and that's not good for a septic tank. Some metal tanks are said to have given twenty years of good service. This might be true, but there are no guarantees that a metal tank is going to last even ten years. In all my years as a contractor, I've never seen a metal septic tank installed. I've dug up old ones, but I've never seen a new one go in the ground.

Fiberglass

I don't have any personal experience with fiberglass septic tanks, but I can see some advantages to them. Their light weight is one nice benefit for anyone working to install the tank. Durability is another strong point in the favor of fiberglass tanks. However, I'm not sure how the tanks perform under the stress of being buried. I assume that their performance is good, but again, I have no first-hand experience.

Wood

Wood seems like a strange material to use for the construction of a septic tank, but I've read that it is used. The wood of choice, as I understand it, is redwood. I guess if you can make hot tubs and spas out of redwood, you can make a septic tank out of it, too. However, I don't think I would be anxious to warranty a septic tank made of wood.

Brick and block

Brick and block have also been used to form septic tanks. When this kind of tank is used, some type of parging and waterproofing must be applied to the interior of the vessel. Personally, I would not feel very comfortable with this type of setup. I don't have any experience with septic tanks made from brick and block, so I can't give you much in the way of case histories.

Installing a simple septic system

Installing a simple septic system is pretty easy if you have the right tools, equipment, and knowledge. As a home builder or general contractor, you might have access to all the tools and equipment needed to make an installation. To illustrate what's involved when adding a septic system, let's run through a typical installation.

The first step in the installation of a septic system is the septic design. This gives you the details needed to make an acceptable installation. The next step is a permit to make the installation. Your local jurisdiction might require a special license to install septic systems, so check on this if you are planning to do your own installations without the aid of an outside contractor.

Your installation begins with the excavation. A backhoe is usually all that is needed for the digging. There needs to be a benchmark set somewhere, probably in association with a tree, that is used for the elevation measurements. These measurements should be precise, so a transit is needed. Once the septic system is laid out on the ground, you are ready to dig. A lot of contractors use ordinary baking flour to mark the dig locations.

Unless there are extenuating circumstances, a standard backhoe is capable of digging the sewer trench, the leach bed, and the hole for the septic tank. It is helpful to have your own dump truck, but you can get by if you have a contract hauler do the work. Dirt and stone can be moved around with the front bucket on the backhoe. The main thing to remember is to follow the septic design to the letter. It is also wise to check your work periodically, to make sure that everything is in keeping with the requirements of the design.

After all excavation is done, you have to install the septic components. The leach field is often done first. With today's modern materials, this work doesn't involve complicated plumbing practices or equipment. A carpenter's saw can be used to cut plastic pipe. Joints are made with a solvent-weld adhesive. As you install the drain pipes, you must refer to the septic design and follow all requirements. Make sure that the septic system is far enough away from the house and well to meet local requirements (Fig. 10-2).

After leach field lines are installed, they are connected to the distribution box. Again, this work is detailed on the septic design. Once the distribution box and field are set up, you are ready to set the septic tank. Keep in mind, the chronological order that I'm giving you is only a suggestion. Some contractors might prefer to set their septic tanks first. It doesn't matter where you start, as long as the job ends up installed properly (Fig. 10-3).

When setting the septic tank, you must make sure it is positioned on solid ground. This is rarely a problem. If the ground is weak, the septic design probably provides some guidance to deal with the problem. It might be necessary to install a layer of stone under the tank,

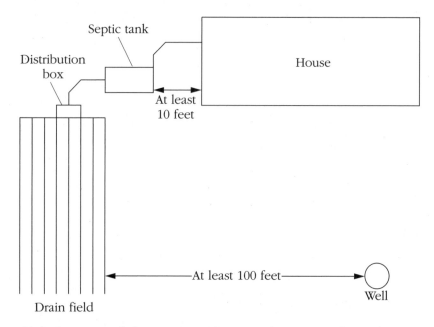

10-2 *Recommended minimum distances between wells and septic systems and septic tanks and homes.*

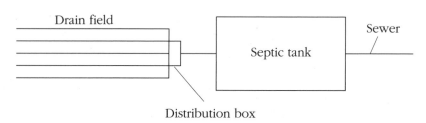

10-3 *Common septic layout.*

but this is not typical. You might have to compact the soil with a tamper, but again, this is not a normal procedure. Usually, a hole is dug and the tank is set into place.

A concrete septic tank is heavy. It is not something that a couple of workers can just horse into a hole. You are probably going to use the front bucket of your backhoe to manipulate the tank. When doing this, make sure that the chain used to lift the tank is strong enough to hold the load. Dropping a concrete tank can damage it, not to mention what it could do to a worker's foot. Don't take any chances when safety is concerned. Don't put yourself or others in a position to get hurt if the chains should slip or snap.

After the tank is in the hole, you probably are going to need to position it with the help of the backhoe bucket and then do some backfilling. Get the tank to sit in place where you want it, then connect it to the distribution box. As a reminder, the pipe used for the run from the tank to the distribution box should be solid, not perforated.

You won't use a glue joint to connect pipe to the septic tank. Instead, the pipe pushes through a precast opening and extends into the tank for several inches. The annular space between the pipe and the opening of the hole in the tank is filled with a cement mixture. You need to make this connection watertight. If the connection isn't watertight, groundwater could run into the tank and sewage could seep out into the ground. Both of these possibilities are undesirable.

An elbow fitting is normally attached to the end of the drain pipe that protrudes into the tank. A short piece of pipe is extended from the elbow into the liquid level of the tank (Fig. 10-4). As the liquid level rises in the tank, it also rises up the short length of pipe and is drained out of the tank into the distribution box.

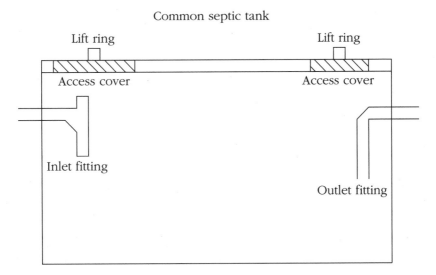

Common septic tank

10-4 *Side view of a septic tank.*

The run of drain pipe from the septic tank to the distribution pipe should be installed with a consistent grade or fall. In other words, the pipe should slope downward, toward the distribution box. A standard grade is ¼ inch of fall for every 12 inches that the pipe travels. For example, a pipe that is eight feet long would be 2 inches higher at the septic tank than it would be at the distribution box. The same grade

is used for the sewer pipe that extends from the building to the septic tank.

Licensed plumbers typically connect building sewers to main
sewers. However, most plumber's limit their rough-in work to a point
not to exceed 5 feet past the foundation. If this is the case, you would
have to run the sewer from the septic tank to within 5 feet of a foundation in order to have a plumber make the connection within the
budget of rough-in work. When a plumber is asked to run sewer pipe
for more than 5 feet, you can expect an extra charge.

Most septic installers run a sewer from the tank to a point near
the foundation. The pipe used for this can be thin-wall sewer pipe.
Many installers prefer to use schedule-40 plastic pipe. This is the
same type of pipe used for drains and vents within a building. You
should consult your local code requirements to determine what options are available to you. Personally, I prefer schedule-40 plastic,
even if thin-wall pipe is an approved material. Long-turn fittings
should be used instead of their short-turn cousins (Fig. 10-5). I recommend installing a clean-out just outside the foundation of the
building (Fig. 10-6).

The sewer that extends between a building and a septic tank
must be installed with a consistent grade. The pipe must be supported by solid ground. It is not acceptable to stick blocks of wood
or rocks under a pipe to support it, and loose fill dirt is not acceptable. The bed of the trench must be solid and graded, so keep this in

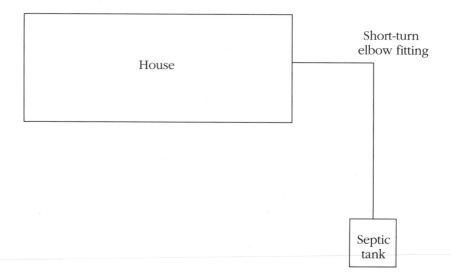

10-5 *Avoid using short-turn fittings between house and septic tank.*

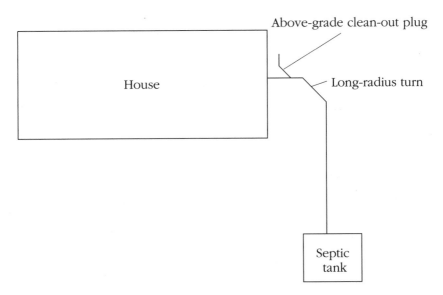

10-6 *Outside cleanout installed in sewer pipe and sweep-type fittings used to avoid pipe stoppages.*

mind as you dig. If your digging produces a trench that has gaps under the sewer pipe, the trench must be filled in to avoid the gaps. This can be done with crushed stone or dirt that can be tamped into place.

The connection of the sewer pipe is parged with cement. So, too is the pipe to the distribution box. Once the sewer pipe is extended into the septic tank, a tee fitting is installed on it. Sewage coming into the tank hits the back of the tee fitting and is directed downward (Fig. 10-7).

Don't cover any of your work until it has been inspected and approved. Every jurisdiction that I've ever worked in required an on-site inspection of septic installations before the system could be buried. Failure to get such an inspection prior to backfilling your work could be very expensive. Make sure the inspection gets done and approved.

Do it yourself

When your next septic job comes along, should you do it yourself? Several things must be considered before you can answer this question. The first one is easy. Can you install your own septic systems legally? Check with your local code enforcement office to see if any

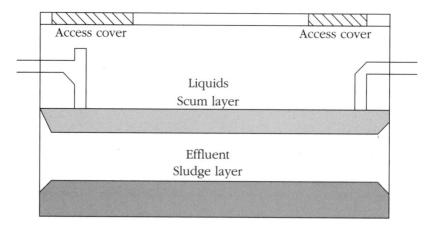

10-7 *Example of the various levels of materials in a septic system.*

special qualifications are needed to install septic systems. You might find that a license of some type is required to obtain septic permits. But, it might be that anyone can install a septic system as long as the installation complies with an approved design. Before you spend any time debating whether or not to install your own systems, get the answer to this question.

Let's assume that you can install your own septic systems. Do you want to? If you don't want to get involved with the hands-on work, you can simply sub the work out to another contractor. This is what I've always done. However, there are some advantages to doing your own installations, and you might wish to consider them.

A builder who installs septic systems can eliminate one subcontractor. Any time you can reduce the number of people who you must count on in the production of a new home, you are one step closer to success. Depending upon other people often gets you in trouble. At least it has always been a sore spot in my business endeavors. I've found that jobs go much better when I have complete control over the work.

Money

Money is almost always a good motivator for expanding your services. If you could make a few thousand dollars extra on a job by installing a septic system yourself, this might be all the reason you need to start doing the work yourself. There is money to be made by installing your own septic systems. The amount of money varies from job to job, but is always enough to make the effort worth considering.

A builder who works primarily with subcontractors might not find it as advantageous to do septic installations. If you don't have your own employees and equipment, it is definitely easier to just sub the work out to a septic installer. You could rent equipment and use your own payroll labor, if you have any, and this should result in some positive cash flow. However, you might find that your time is better spent selling new jobs than it is overseeing the installation of a septic system. I've often found myself in that position.

I've never owned heavy equipment. Piece workers and subcontractors have always made up most of my work force. While I believe in having some employees available, my building business centers around subcontractors much more than it does employees. For these reasons, it has never seemed cost-effective for me to get into doing my own septic installations.

My situation is a little unusual in that I'm a licensed builder and a licensed master plumber. The fact that I maintain both a building and a plumbing company makes me an ideal candidate to create a septic crew. Most of my work is presently done in areas where public sewers are not available. Even with all of this going for me, I've still never jumped into the septic business.

One reason I've probably not made a transition into septic work is the need for equipment and operators. My volume of septic work would not support the purchase and upkeep of a backhoe, trailer, and dump truck. Neither would it keep an operator and driver busy. Come to think of it, this is probably why I haven't pursued the opportunity. If I had the equipment and personnel, I'm quite sure I would do my own septic installations.

If I were looking for an expansion option for my existing businesses, septic work might be a consideration. However, site contractors often have a lock on septic work. Since these contractors bid clearing, driveway installations, grading, excavation, and so forth, they are a natural source for septic installations. This, too, is probably another reason why I haven't made financial commitments to get into the septic business.

Work is work

When times are tough, work is work. Installing your own septic systems can put some extra money in your pockets. It can also help to keep your crews busy. This is a plausible reason to consider doing your own installations, even if you have to rent equipment. If you have a choice of sending your crews home for lack of work, having them push inventory around for something to do, or installing a septic system, you should come out ahead by having them put in the septic system.

Control and quality

Control and quality are both good reasons for doing your own septic installations. If your crews do the job, you have more control over the work than when its done by a subcontractor. It also stands to reason that you have better control over the quality of the work being done. Outside of money, this might be the best reason to do your own septic work.

The technical side

The technical side of installing septic systems is not difficult to understand. If you can read blueprints, you should be able to interpret a septic design. Once this goal is accomplished, there is very little to stop you from installing septic systems properly. Field experience helps, as it does in any type of work, but installing a septic system is not a job that requires a lot of hands-on experience. I don't wish to minimize the skill required for septic work, but compared to other types of trade-related work, it is pretty easy to understand.

Bad ground

Bad ground can force you away from a simple pipe-and-gravel septic system. If the percolation rate is not sufficient, you are going to have to look to some other type of system, such as a chamber system. These expensive systems are discussed in the next chapter.

Some soils are not absorbent enough to accommodate a typical pipe-and-gravel septic system. You must be sure of your septic options before you bid a job. If you typically plug in a generic figure to represent the price of a septic system, without first reviewing the septic design, you are setting yourself up for big trouble. A day will come when you are forced to use a chamber or pump system, and your typical price is going to be way too low.

We could talk about trench and mound systems here, but I would prefer to save them for the next chapter. Trench systems are often similar to pipe-and-gravel systems, but they are not as common. Therefore, I'd prefer to talk about them with mound and chamber systems. If you're ready, let's move to the next chapter and examine additional septic options.

11

Chamber-type septic systems and other special-use systems

If you are dealing with a job where a typical pipe-and-gravel septic system can't be used, you are faced with installing a chamber-type or other special-use system. Any of these systems typically cost more to install than a simple pipe-and-gravel system. The specific type of system used depends greatly on the existing soil conditions. Choosing a special-use system is rarely left to the discretion of a builder. Engineers and county officials normally dictate the type of special system that must be used.

There are four common types of special-use septic systems. Chamber systems are the most common. Trench systems are usually the most inexpensive type of special-use system available. A mound system might be prescribed, and there are times when a pump system is required. Your installation cost with any of these systems is likely to be higher than what you might expect.

Chamber systems are very common in Maine. I'd never encountered a chamber system until I moved north. The soil in Virginia tended to be well suited to septic systems, so gravel-and-pipe systems were normally installed. Pump systems were required occasionally, but chamber systems were never required on any of my jobs in Virginia.

My personal experience with septic systems is limited to the East Coast. I don't have first-hand experience with septic systems in other parts of the country. However, a little research on your part can reveal the types of systems that are the most common in your area. Since septic systems require septic designs and permits to be installed, it is fairly easy to research the types of systems that are used in your region.

Chamber systems

Chamber septic systems are used most often when the perk rate is low. Soil with a rapid absorption rate can support a standard pipe-and-gravel septic system. Clay and other types of soil might not. When bedrock is close to the ground surface, as is the case in much of Maine, chambers are often used.

What is a chamber system? A chamber system is installed very much like a pipe-and-gravel system, except for the use of chambers (Fig. 11-1). The chambers might be made of concrete or plastic. Concrete chambers are more expensive to install. Plastic chambers are shipped in halves and put together in the field. Since plastic is a very durable material, and it's relatively cheap, plastic chambers are more popular than concrete chambers.

Chamber systems typically require many chambers. The chambers are installed in the leach field, between sections of pipe. As effluent is released from a septic tank, it is sent into the chambers. The

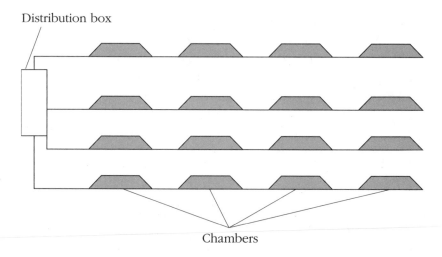

Distribution box

Chambers

11-1 *Example of a chamber-type septic field.*

chambers collect and hold the effluent for a period of time. Gradually, the liquid is released into the leach field and absorbed by the earth. The primary role of the chambers is to retard the distribution rate of the effluent.

Building a chamber system allows you to take advantage of land that would not be buildable upon with a standard pipe-and-gravel system. Based on this, chamber systems are good. However, when you look at the price tag of a chamber system, you might need a few moments to catch your breath. I've seen a number of quotes for these systems that pushed the $10,000 mark. Ten grand is a lot of money for a septic system. But, if you don't have any choice, what are you going to do?

A chamber system is simple enough in its design. Liquid leaves a septic tank and enters the first chamber. As more liquid is released from the septic tank, it is transferred into additional chambers that are farther downstream. This process continues with the chambers releasing a predetermined amount of liquid into the soil as time goes on. The process allows more time for bacterial action to attack raw sewage, and it controls the flow of liquid into the ground.

If a perforated-pipe system is used in the ground where a chamber system is recommended, the result could be a flooded leach field. This might create health risks. It would most likely produce unpleasant odors, and it might even shorten the life of the septic field.

Chambers are installed between sections of pipe within the drain field. The chambers are then covered with soil. The finished system is not visible above ground. All of the action takes place below grade. The only real downside to a chamber system is the cost.

Trench systems

Trench systems are the least expensive version of special septic systems. They are comparable in many ways to a standard pipe-and-gravel bed system. The main difference between a trench system and a bed system is that the drain lines in a trench system are separated by a physical barrier. Pipe-and-gravel bed systems consist of drain pipes situated in a rock bed. All of the pipes are in one large bed. Trench fields depend on separation to work properly. To expand on this, let me give you some technical information.

A typical trench system is set into trenches that are between 1 to 5 feet deep. The width of the trench tends to run from 1 to 3 feet. Perforated pipe is placed in these trenches on a six-inch bed of crushed stone. A second layer of stone is placed on top of the drain pipe. This rock is covered with a barrier of some type to protect it from the backfilling process. The type of barrier to be used is specified in a septic design.

When a trench system is used, both the sides of the trench and the bottom of the excavation provide an outlet for liquid. Only one pipe is placed in each trench. These two factors are what separates a trench system from a standard bed system. Bed systems have all of the drain pipes in one large excavation. In a pipe-and-gravel bed system, the bottom of the bed is the only significant infiltrative surface. Since trench systems use both the bottoms and sides of trenches as infiltrative surfaces, more absorption is possible.

Bed or trench systems should not be used in soils where the percolation rate is either very fast or slow. For example, if the soil accepts 1 inch of liquid per minute, it is too fast for a standard absorption system. This can be overcome by lining the infiltrative surface with a thick layer (about two feet or more) of sandy loam soil. Conversely, land that drains at a rate of 1 inch an hour is too slow for a bed or trench system. This is a situation where a chamber system might be recommended as an alternative.

More land area

Because of their design, trench systems require more land area than pipe-and-gravel bed systems. This can be a problem on small building lots. It can also add to the expense of clearing land for a septic field. However, trench systems are normally considered to be better than bed systems. There are many reasons for this.

Trench systems are said to offer up to five times more side area for infiltration to take place. This is based on a trench system with a bottom area identical to a bed system. The difference is in the depth and separation of the trenches. Experts like trench systems because digging equipment can straddle the trench locations during excavation. This reduces damage to the bottom soil and improves performance. In a bed system, equipment must operate within the bed, compacting soil and reducing efficiency.

A trench system is ideal for hilly land. The trenches can be dug to follow the contour of the land. This gives you maximum utilization of the sloping ground. Infiltrative surfaces are maintained while excessive excavation is eliminated.

The advantages of a trench system are numerous. For example, trenches can be run between trees. This reduces clearing costs and allows trees to remain for shade and aesthetic purposes. However, roots might still be a consideration. Most people agree that a trench system performs better than a bed system. When you combine performance with the many other advantages of a trench system, you might want to consider trenching your next septic system. It costs more to dig individual trenches than it does to create a group bed, but the benefits might outweigh the costs.

Mound systems

Mound systems, as you might suspect, are septic systems that are constructed in mounds that rise above the natural topography (Fig. 11-2). This system compensates for high water tables and soils with slow absorption rates. Due to the amount of fill material needed to create a mound, the cost is higher than it would be for a pipe-and-gravel bed system.

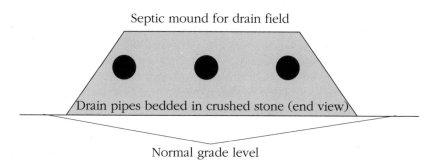

Septic mound for drain field

Drain pipes bedded in crushed stone (end view)

Normal grade level

11-2 *Cut-away of a mound-type septic system.*

Coarse gravel is normally used to build a septic mound. The stone is piled on top of the existing ground. However, topsoil is removed before the stone is installed. When a mound is built, it contains suitable fill material, an absorption area, a distribution network, a cap, and topsoil. Due to the raised height, a mound system depends on either pumping or siphonic action to work properly. Essentially, effluent is either pumped or siphoned into the distribution network.

Treatment of the wastewater occurs as the effluent passes through the coarse gravel and infiltrates the fill material. This continues as the liquid passes through the unsaturated zone of the natural soil.

The purpose of the cap is to retard frost action, deflect precipitation, and to retain moisture in order to stimulate the growth of ground cover. Without adequate ground cover, erosion can be a problem. A multitude of acceptable ground cover choices are available. Grass is the most common choice.

Mounds should be used only in areas that drain well. The topography can be level or slightly sloping. The amount of slope that's allowable depends on the perk rate. For example, soil that perks at a rate of 1 inch every 60 minutes or less, should not have a slope of more than 6 percent if a mound system is to be installed. If the soil absorbs water from a perk test faster than 1 inch in one hour, the slope could be increased to 12 percent. These numbers are only examples. A professional who designs a mound system can set the true criteria for slope values.

Ideally, about 2 feet of unsaturated soil should exist between the original soil surface and the seasonally saturated top soil. There should be 3 to 5 feet of depth to the impermeable barrier. An overall range of perk rate could go as high as 1 inch in two hours, but this, of course, is subject to local approval. Perk tests for this type of system are best when done at a depth of about 20 inches. However, they can be performed at shallow depths of only 12 inches. Again, you must consult and follow local requirements.

The design and construction of mound systems can get quite complicated. This is not a problem to you, as a builder, because experts provide the design criteria. It's then up to you or your septic installer to follow instructions and see to it that the mound is built as specified.

Pump systems

We are going to have a detailed discussion about pump systems in Chapter 13. However, since pump systems are special systems, they do deserve mention in this chapter.

Pump systems are expensive. They are needed when a septic system is installed at an elevation that is higher than the sewer. This situation is not as uncommon as you might think, especially when dealing with houses that have plumbing in the basements. However, houses with basements are not the only structures that might need pumping systems.

If you are bidding a job that requires a pumping system, you must be careful not to overlook the pump and related needs that are to be installed. Additional labor and material costs for pump systems can be very substantial. For more information on pump systems, refer to Chapter 13.

Holding tanks

Holding tanks are sometimes used as a last resort for building sites that have very poor perk rates. The use of holding tanks might or might not be acceptable in your area. Don't buy land on the assumption that you can install holding tanks as an alternative to an absorption septic system. Even if holding tanks are allowed in your area, the use of such a waste disposal system is likely to have a negative impact on the value of any building you build.

Some land simply isn't suitable for any reasonable type of absorption septic system. This typically labels the land as unbuildable. Under these conditions, holding tanks might be a solution. Basically, a holding tank is just a large vessel that collects wastewater and sewage. The container holds the waste until a certain level is reached in the tank. At this point, the tank must be emptied. This is done with a truck that is set up to pump out septic tanks. How often the holding tank must be pumped depends upon the size of the tank and the volume of use.

Pumping out holding tanks gets expensive after awhile. The process is also something of an inconvenience. Some people are willing to put up with the hassle and expense, but most are not. If you are building on spec, avoid sites that require holding tanks unless you have a very good plan of attack for selling what you build. Most home buyers are not going to like the idea of having a holding tank rather than a septic system.

I can think of very few occasions when holding tanks have been used for full-time residences. This septic alternative is better suited to seasonal cottages and camps. However, you might find an occasion when a holding tank can help you out of a tight spot. Remember, check with your local code office to see if holding tanks can be used before you count on them as a solution to your septic system problems.

Other types of systems

Engineers might be able to offer other types of systems to meet your special needs. It might be that a combination of systems can be pieced together to accomplish a difficult goal. If you run across a piece of land that doesn't perk well, avoid it if you can. Unless you are able to buy the land very inexpensively, as an offset to the poor perk rate, the cost of a special septic system could ruin any hope of profitably building on the land. You must weigh each individual set of circumstances to determine what is worth pursuing. When you are asked to custom-build something for a customer on land that exhibits difficult septic characteristics, have the customer consult engineers or other professionals for a detailed septic design. As long as you are working from an approved septic design, you should be able to keep yourself out of trouble.

I told you that we would talk more about pump systems in Chapter 13. Before we do that, I'd like to discuss gravity systems. Gravity systems are less expensive and more desirable than pump systems. To learn more about gravity systems, let's turn the page and get into Chapter 12.

12

Gravity septic systems

When it comes to desirability and cost-effectiveness, gravity septic systems reign supreme. Not only are gravity systems the most common type installed, they are also the most dependable and least expensive.

Unlike pump systems, gravity systems don't require any routine mechanical care. A gravity system works without electricity. Since a pump station isn't needed with a gravity system, users don't have to worry about flooding the holding tank. Gravity systems don't have a holding tank. The only maintenance required is the periodic pumping of the septic tank. For most homes, the septic tank has to be pumped only once every couple of years.

A gravity system can save thousands of dollars in construction costs. The peace of mind that comes with a gravity system is also comforting. Users don't have to wonder if the sewage pump is going to fail at an inopportune time. Gravity systems don't need a pump. Basin floats don't stick with gravity systems, because they don't exist. False alarms don't sound at a control panel when using a gravity system. Why? Because there is no need for a control panel to monitor liquid levels. The advantages of a gravity system are numerous, to say the least.

Gravity systems can't serve all homes. General topography can force a builder to use a pump system. The house design can play a part in determining what type of septic system is going to be needed. Builder planning can have a lot to do with whether or not a gravity system can be used. With some experience and preplanning, a builder can often avoid the use of a pump system.

To avoid a pump system, there are several good reasons why you should pay close attention to the planning of a house. Cost is one of them. Customer satisfaction is another. Also, there is a reduced risk of call-backs and warranty work. The list could go on, but it is enough to say that gravity systems are, by far, the best choice when given an option.

Luck of the draw

Does the luck of the draw decide whether you install a gravity system or a pump system? No, it doesn't. Sometimes a pump system is the only viable option. But, on many occasions, the decision could go either way depending upon the steps taken and the planning that's done. As a builder, can you influence the type of system that is going to be used? Yes, you can. By talking with your septic installer and plumber, you might be able to turn the odds of a close call in your favor. Let me give you a few examples.

House placement

The placement of the house on a building lot can make the difference between a gravity system and a pump system. Moving a house a few feet in one direction or another might be all that is needed to switch from a pump system to a gravity system. This was the case with my most recent personal home.

My house is located on acreage, so lot size was not a problem when laying out the well, septic system, and house site. However, there was one place in particular that I wanted to put the house. The site I chose put a pine forest outside my sitting room window, a grove of white birch trees just outside the breakfast room and kitchen, and river frontage in front of the living room, dining room, sunroom, and master suite. Part of my motivation for buying that property was the river and trees. It was important to me to position the house to take advantage of the available views.

After having my soils tested, I was cleared for a pipe-and-gravel septic system. This actually surprised me. I was expecting a chamber system, due to the presence of bedrock. Getting the opportunity to save some money never hurts my feelings, so I was thrilled to learn of my good fortune. However, I noticed on the septic design that I might be required to install a pump station for the septic system. This didn't please me.

Septic designs should be done long before any foundation work has begun. This being the case, I had ample notice of a potential need for a pump system. Wanting to avoid a pump at all reasonable costs, I called a meeting with my septic installer. There was no need to call in a plumber, since I was doing all the plumbing work. However, a typical builder who is not a master plumber would be wise to also solicit the advice of the plumbing contractor.

My septic installer and I looked over the septic design and the site plan. There were two reasons why a pump system might be needed. Bedrock was close to the surface, so there was concern that it would be difficult to bury the septic tank. If the tank could not be put deep enough into the ground, a pump would be needed. The other concern dealt with the house placement. Due to the slope on my land at one corner of the house, it appeared as if the sewer might be lower than the septic tank. This would, of course, require the use of a pump.

Like many homeowners, I was set in my ways about where I wanted the house situated on the land. But I was not opposed to moving it a few feet forward or backward to avoid the use of a septic pump. I was faced with two potential problems to solve if I wanted to eliminate the need for a septic pump. Let me tell you how I solved them.

My biggest obstacle to overcome was the bedrock. I didn't want to blast it out of the way, so I was forced to work around it. Since the land sloped at one corner of my proposed house site, I reasoned that we could put the tank into the side of the slope and fill in around it with dirt that would be excavated for the foundation. My septic installer used his transit to check elevations from the benchmark and found that my idea would work, assuming that I could get the sewer from the house to the tank with an adequate grade.

Due to bedrock, I opted for a crawl-space foundation. The only way to get a full basement would be to blast out the rock, and this was far too expensive for me to seriously consider. The plumbing code requires a minimum of 12 inches of ground cover over a sewer pipe. After working with my blueprints for awhile, I devised a way to get my sewer out high enough to run on grade (at ¼ inch to 1 foot) to the septic tank. To do this, I had to shift the house location by about five feet, but this didn't affect the overall views I wanted to achieve, so I authorized the move.

Once the septic tank location was moved and the house foundation was relocated, it looked like a pump would not be needed. To be sure, I had the soils engineer check my plans. He concurred that the plan was viable and would eliminate the pump system. By spending a few hours at the drawing board, I was able to eliminate the extra cost of a pump system and the potential long-term headaches that go with such a system. This is a good example of how a little planning can save you a lot of money and trouble.

Let me give you another example from my past. When I was contracted to build a house some time ago, the customer told me that a pump system would be needed for the septic field. Because a bathroom was scheduled to be installed in the basement, it was impossible to bring the sewer out above the septic tank. I could have accepted this and gone on about my business, but I decided to voice my opinion.

I recommended that the main sewer be located to run out the foundation wall at a height that would eliminate the need for a pump station. The homeowner thought this would mean giving up the basement bath. Then I explained that a basin and sewer pump could be installed to serve the basement bath, and that the rest of the house could work off of a gravity system. Can you see the advantage to this change in plans?

When a main sewer dumps into a pump station, the cost of a septic system skyrockets. There is the cost of the holding tank, the pump, and the alarm controls. If the system fails, the whole house is shut down in terms of plumbing drainage. By installing a pump system for only the basement bath, a lot of money can be saved, and the rest of the house is not dependent on a pump for sewage removal.

It isn't a big deal to install a pump basin and pump in a basement. Of course, the cost is more than if it weren't needed, but not nearly as expensive as a full-scale pump station. Since it is used only for the basement bath, the pump wear and tear is reduced. Also reduced is the possibility of a total plumbing failure. Waste from the basement bath is pumped up to the main building drain, inside the house, and then runs out to the septic tank under a gravity feed. Such a plan can often be used to your benefit.

In the example I've just given you, the customer wanted me to build the house. I was not in a competitive bid situation. But, if I had been, do you think my suggestion would have given me a competitive edge? Of course it would have. I would have had an advantage in two ways. First, my price would have been lower than contractors who were bidding the job with a full-scale pump station. Secondly, my recommendation would have shown my concern for the customer and my expertise in building alternatives. This would have built confidence in the customer, giving me an edge that couldn't have a price tag put on it.

Put yourself in the place of a homeowner who feels that a pump system is the only choice for a septic system. If you met a contractor who offered you alternatives, wouldn't you be impressed? When you saw that you could save thousands of dollars at the time of construction and eliminate the risk of a total plumbing failure and ongoing maintenance costs, you should be impressed even more. This type of bidding strategy can be all it takes to win jobs that you might otherwise lose.

Luck has very little to do with whether or not a pump system is required. Evaluation of land is the first key to determining the type of septic system that is needed. House placement is another element in the conclusion. Creative options also play a vital role in avoiding pump stations. Experienced builders have the opportunity to control all of these factors. Therefore, you can have a lot to do with whether or not a pump station is going to be installed on your next septic job.

Look out

If you know what to look out for when bidding a job or evaluating land, you can reduce the odds of having to install a pump station. Both the land and the house plan have a bearing on whether or not a pump system is needed. Since land acquisition must be determined before an accurate building estimate can be given, let's start with land evaluation.

Land evaluation

Land evaluation means a lot to a builder. Until you see a building site, you cannot realistically work up an accurate cost of building a house. You have to know what the site conditions are to figure prices for site work, foundation work, and other aspects of a job. You must also know the site conditions to speculate on the type and cost of a septic system.

You should not attempt to bid on the cost of a septic system until you have an approved septic design. This doesn't mean that you can't get a fair idea of what to expect by looking at the land. We've already talked about soil qualities in terms of bed systems, trench systems, mound systems, and chamber systems. Our goal at this point is to identify items to look for that can influence the need for a pump system.

Can you think of anything in particular that you should be on the lookout for? Well, bedrock is one risk. If you can't dig deeply enough to bury a septic tank below the sewer it serves, you are going to have to install a pump system. One alternative might be blasting, but the cost of this type of work is scary. How can you tell if bedrock is going to be a problem? Sometimes you can see bedrock. A stiff, metal rod can be used to probe for bedrock. To get the rod deep enough for a good test, you probably have to drive it into the ground with a heavy hammer. I've said it before, but I'll say it again, make sure that there are no underground utilities in the area where you are probing.

If rock is not a factor, water could be. A high water table can limit the depth at which a septic tank can be buried. Short of installing underground piping to divert groundwater, holding a septic tank up higher is the only alternative. And, this means that a pump is more likely to be needed. How can you test for groundwater? A hand-held auger works well, and posthole diggers can also get the job done. By digging or boring some test holes, you can see if water is likely to be a problem.

Topography is often a major factor when determining the need for a pump station. If the only land that perks well enough for a septic field is higher than the house site or sewer, you have no choice but to go with a pump. You might be able to move the house, but beyond this, your options are nil.

Some land has such obvious differences in elevation that an instrument is not needed to check the need for a pump system. Many building lots don't exhibit such extreme changes in elevations. Under these conditions, a transit is very helpful for determining the feasibility of a gravity system. Even some string and a line level can be used to pull figures for elevation comparisons. Optical illusions do exist, and they can cost you a lot of money if you think a gravity system can work and then find out later that a pump is needed. Don't bid a job until you are sure of what you are bidding.

House factors

What house factors can influence the type of septic system used? The location of plumbing fixtures has a direct effect on the need for a pump. If fixtures are installed below the level of a sewer or septic tank, a pump is going to be needed. This does not, however, necessarily mean that a full-scale pump station is needed. I already gave you an example of how a sump basin and grinder pump can serve a basement bath and avoid a whole-house pump station. A similar set-up can be installed to pump the discharge from a clothes washer up to a gravity-fed system.

When the only fixture below a gravity system is a sink, you can have your plumber install a small sink pump. This is a device that attaches right to the bottom of the sink and pumps wastewater up to a gravity system. It would be ridiculous to install a whole-house pump station to accommodate a clothes washer or sink that is below the gravity line.

Foundation height and the finished grade level of a lawn can both be factors in the need for a pump. Foundations can be held a little

high to allow sewers to exist on a gravity basis. Due to code requirements on the coverage of pipes, you must allow room for ample grading where the sewer leaves the house. This can usually be worked out
without much problem.

When you are evaluating a house for a gravity system, you must
pay attention to where the sewer leaves the house. Could you avoid
a pump system by routing the plumbing in a different way? You often can. Distance is your enemy with a gravity system. The farther a
pipe must run, the more grade it needs. When a long sewer run is
needed, the septic tank must be deeper. However, you might be able
to avoid a pump if you can shorten the run by bringing the pipe out
at an opposite corner.

Talk to your plumber about altering the minimum pipe grade.
One way of doing this is by using a larger pipe. The minimum amount
of grade required by plumbing codes for a sewer is tied to the pipe
size. A 4-inch pipe is normally used as a sewer. You can use 3-inch
pipes if no more than two .oilets discharge into the pipes. Larger
pipes can always be used. A quick check of your local plumbing code
can give you the information needed on minimum grade requirements. Most plumbers like to work with a grade of ¼ inch per foot.
But, with a 4-inch pipe, a grade of ⅛ inch per foot is normally acceptable. This might not seem like a big difference, but those fractions of
an inch can add up in a long sewer run.

Moving a septic tank

Moving a septic tank closer to a house can improve your odds of
eliminating a pump. When you position a septic tank close to a
house, you minimize the grade needed to get the sewer to the tank
on a gravity basis. Regulations dictate just how close a septic tank can
be from a foundation. A common setback requirement is 10 feet, but
check your local regulations to find out the rules in your region.

A second opinion

Sometimes it pays to get a second opinion. This new opinion might
come from the same soils expert who designed the system in the first
place, or it might come from a different professional. If you have a
septic design that calls for an expensive pump system, you might be
able to eliminate it simply by finding another location for the leach
field. A call to an engineer for a second opinion might be all it takes.
Let's look at an example of how this might work.

Assume that you are a builder who develops land into building
lots. It is common practice for a county employee to perform site

evaluations and perk tests so that septic designs can be created. You jump through the normal hoops and have the county inspect the property you are planning to buy. Unfortunately, many of the sites call for pump stations. Do you have any options? Yes, you can seek a second opinion from an engineering firm.

If you go to an engineering company and request soil tests on the parcels that require pump systems, you are going to have to spend some out-of-pocket cash. But, the money might be well spent. An engineer working under your directions might work a little harder to find sites suitable for gravity systems. Now don't get me wrong. Sometimes a pump system is the only alternative. But, there are many times when the need for a pump is marginal, and this is when you can do the most to avoid a pump system. If your engineer can find a suitable site other than the one found by the county employee, you're all set. I doubt if there is any jurisdiction that has not acquiesced to the findings of suitable professionals, such as engineers, from time to time.

Code officers are not monsters who are out to make your life miserable. They have a job to do, and they do it. If you can provide proof to a code officer that an alternative method is safe and meets the intent of the code, you are likely to gain approval. Engineers are good at making these cases. The cost of paying for independent soil studies is a burden, but it could make your next project much more desirable and more profitable. If the stakes are high enough to warrant the cost, don't overlook the option of getting a second opinion.

Don't accept the obvious

Contractors who are good at avoiding pump systems don't accept the obvious. If you are handed a septic design that calls for a pump station, the easy way out is to accept it and bid the job. I don't recommend this. Technically, there is nothing wrong with bidding a job as it is presented, but going to a little extra trouble can make all the difference in the world in winning the bid. If you simply accept the fact that a pump is needed, you fall into a category with most other contractors. Showing a little creativity by seeking an alternative method of installing the septic system could set you apart from the competition and assure your success in winning the bid.

If you are a spec builder, you simply can't afford to accept the fact that a pump system is required for the house you are about to build. Pump systems cost much more than gravity systems, and this increase in cost must be reflected in the sales price of your house. Raising the price of your house might mean that it takes longer to sell. If this is the case, interest on your construction loan adds up, eating into your projected profits. Qualified home buyers are grouped into price ranges. Raising the price of a spec house by $3000 could move you into a new category of buyers, eliminating an entire group that might have bought the house at a lower price. Spec builders cannot afford to alienate any potential buyers.

A house that relies on a pump to drain its plumbing fixtures is not as desirable as one that doesn't need a pump. This is another strike against the salability of your product. Building on speculation is tough enough under the best of conditions, you certainly don't need to add to this burden by stacking the cards against yourself.

I've been a builder for a long time, and I know that people shy away from overpriced houses and houses that have pump stations for their septic systems. In this chapter, I've given you many ways to avoid pump stations. Your choices are clear. You can roll over and accept the first ruling you get that requires a pump station, or you can fight to eliminate the need for the pump. What are you going to do? I'd fight.

If you are a land developer, as I have been, the issue of suitable perk tests becomes even more important. A developer selling lots that require pump-style septic systems is at a great disadvantage when competing with developers who have land that needs only gravity systems. I can't stress enough how important it is to obtain permission for a gravity septic system.

Have I ever installed a pump system for one of the houses I've built? Yes, but only once. In all other instances, I've been able to find a way to use gravity systems. This is not to say that I haven't been dealt my share of pump-system lots. I just didn't give up until I found a way of getting around the problem. On the one occasion when I was forced into a pump system, it was either take the pump system or lose the job. The building lot was such that no alternative was available.

Would I ever develop land with the knowledge that pump systems would be required? Probably not. Unless I was able to acquire the land at some outrageously low price, I would stay away. As you might have guessed, I'm not a fan of pump systems. I know them well as a plumbing contractor, and I wish to avoid them whenever possible as a builder.

The next chapter is dedicated to pump systems. As much as I dislike them, they are sometimes necessary. I can tell you a lot about these systems, both as a builder and as a plumber. So, let's turn the page and see what you are in for if you are dealing with a pump system.

13

Pump stations for difficult situations

Pump stations are sometimes the only answer when you have difficult septic situations. We have talked about avoiding pump systems whenever possible, but sometimes a suitable alternative is just not available. When that happens, a pump system must be installed. It's not that pump stations are difficult to install, it is the cost that makes them hard on a builder. The routine maintenance and potential for failure, along with the cost, makes pump systems tough for the property owner. But, if a pump system is the only way to have a septic system, you don't have much choice.

If you build in rural areas long enough, you are going to eventually come across a piece of land that needs a pump station for the septic system. I've only installed one whole-house pump station during my building career. But, I've installed countless pump stations for plumbing fixtures located in basements, garages, and similar locations. It is not always necessary to pump the discharge of all the fixtures in a house. Sometimes only one fixture needs to be pumped. It might be an entire fixture group that has to be pumped, such as a full bathroom. If at all possible, try to avoid installing a whole-house pump station.

We are going to discuss three types of drainage pumping in this chapter: single-fixture pumps, basin pumps that are located within a house, and whole-house pump stations. Any of these methods might solve your problems. Some are cheaper and easier to install than others. All of them add some cost to a job, so you must be aware of this during the bidding phase.

Single-fixture pumps

Single-fixture pumps are inexpensive and easy to install. They don't have any real effect on plumbing except for the fixtures on which they are installed. This type of pump is often used on laundry tubs. In fact, a lot of plumbers call them laundry-tray pumps. The pumps can, however, be installed on any type of residential sink or lavatory. They are not suitable for bathtubs, showers, toilets, or clothes washing machines.

Actually, there is one occasion when single-fixture pumps are suitable for use with clothes washers. If the indirect waste of the washing machine dumps into a laundry tub, the pump can be used to empty the contents of the tub. Even though the pump is limited in its flow rate, the holding capacity of a deep laundry tub is enough to allow the pump to keep up with the volume of water discharged by an automatic clothes washer.

A single-fixture pump is small. It installs directly under the bottom of the fixture it is serving. Electricity is needed to make the pumps run. An experienced plumber can add a single-fixture pump to a new installation in just a matter of minutes. You must also take into consideration, however, the expense of the pump and the electrical circuit that is needed. All of these factors tend to push up the price of your job.

If you have only one sink that must be installed below the gravity level of other plumbing, a single-fixture pump is a cheap, easy way out. You often find them on basement bar sinks. The discharge pipe from a single-fixture pump normally has only a ¾-inch-diameter drain. This makes it easier to pipe the drain in concealed locations.

Small, single-fixture pumps have some disadvantages. Depending upon the type of fixture being served, the pumps can become clogged with sediment and debris. This is often the case when they are used with laundry tubs. When a clothes washer empties into a laundry tub, it often deposits lint in the sink. This lint, if it gets into the drainage system, can plug up the strainer in a small pump. One way to avoid this is by using a strainer to protect the drain opening of the laundry tub. But, the strainer could clog and cause the sink to overflow. You could be fighting a losing battle either way.

Single-fixture pumps can be adjusted to cut on as soon as water is detected in a drain, or they can be set to wait until a specific amount of water pressure builds up. The pumps last longer if they don't have to cut on and off frequently and for short durations. Therefore, a setting that allows water to collect in a plumbing fixture before being pumped out is advantageous. It is also possible to use the pumps in a manual fashion, where someone turns the pump on and off as needed. This, of course, would not be suitable in the case of draining an automatic clothes washer, but it can work well for other applications.

The strainers in single-fixture pumps are sensitive. Dirt, sand, lint, and other items are all capable of blocking the pump filter, causing a back-up of water in a fixture. Most pump failures that I've witnessed have come when the pumps were used on laundry tubs with nothing more than cross-bar protection over the drain. Pumps installed on lavatories and bar sinks don't get the sediment and debris that laundry trays get, so they are less likely to fail.

While single-fixture pumps are inexpensive and effective, they are not always the best choice. I've just explained to you how the small pumps can become clogged and cause back-ups in fixtures. This type of failure could result in flooding. For example, if the single-fixture pump for a laundry tub failed, the discharging washing machine water could overflow.

How often do single-fixture pumps fail? Provided that they are protected from contaminants and are not abused, they perform very well. My plumbing company has responded to a lot of calls for failed pumps on single fixtures, but in most cases, the failure was related to misuse. By this, I mean that people allowed objects to go down a drain that were never intended to pass through a small pump.

I would not hesitate to install a single-fixture pump in my own home, but I'm always a little nervous when I install them for other people. Since I don't know how my customers treat the pumps, I could end up getting frantic calls from people who are faced with a house that is flooding.

When I install a single-fixture pump, I have a little instructional sheet that I give my customers. It warns them of the risks associated with careless use of the pump. Before I leave a job, I ask the customers to read and sign the form. I give a copy to them, and I keep a copy for my records. This gives written documentation that I instructed the individuals in the proper use of their pumps. If a problem arises from abuse or negligence on the part of a homeowner, my little piece of paper helps give me some protection.

Most plumbers don't go to the extremes that I do in order to get paperwork signed for a pump installation. In all my years of plumbing and building, only once have I had to produce my form as a defense. This was after two free service calls to clear blockages in a strainer of a pump. On both occasions, the pump strainer was full of lint and sand. After the second trip, I produced the paperwork and instructed the homeowner that I would have to charge them for future service calls if sand or lint was found in the strainer. Since then, the customer has never called me back with a pump problem.

There is an option to single-fixture pumps when only one fixture needs to be pumped. A more costly, but better, pumping arrangement is available for all types of plumbing fixtures. Depending on the type and number of fixtures that need to be pumped, requirements for this alternative vary. Let me explain by breaking the situations down into individual examples.

Gray-water sumps

Gray-water sumps and pumps can be used to collect and pump water from all plumbing fixtures that don't receive or discharge human waste. In a typical residence, this would include all fixtures except toilets. Many commercial buildings use gray-water sumps to handle wastewater from sinks, and a lot of gray-water sumps are installed in houses for one reason or another.

What is a gray-water sump? The sump itself is not much more than a covered bucket (Fig. 13-1). It is simply a watertight container that is used to collect wastewater from a plumbing fixture. The sump can be installed by sitting it on the floor under a fixture, or it can be

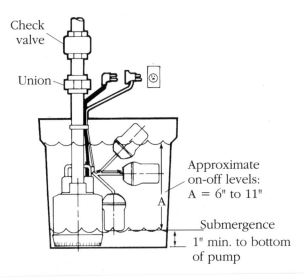

13-1 *Submersible sump pump with a standard sump installation.* Goulds Pumps, Inc.

buried in a concrete floor. Burying the sump or suspending it beneath a floor would be necessary if it were to serve a bathtub or shower.

A drain is run from a plumbing fixture to the sump, where a watertight connection is made. Another pipe is connected to the discharge outlet at the sump and is run to a drain pipe that empties by gravity. The pipe used for this type of pump often has a diameter of 1½ inches, although a larger pipe could be used.

The cover of the sump is often fitted with a gasket that comes into contact with the rim of the sump when the cover is installed. A sump cover is normally held in place with machine screws. An air vent should extend from the top of the sump to either open air space or to a connection with another plumbing vent that terminates into open air. The diameter of the vent pipe is usually 2 inches.

Gray-water sumps normally have a holding capacity of about 5 gallons. The sumps that I use are made of polyethylene, so there is no need to worry about corrosion. When equipped with a gas-tight cover and vent, a gray-water sump does not emit odors. A key part of a sump system is the pump (Fig. 13-2). Many pump options are available. Pumps with built-in check valves are my first choice.

I buy a kit when I install a gray-water sump system. The kit includes the sump and a pump. The pump is a one-third horsepower unit that requires 115-volt electrical service. The amp draw is 2.3, and the pump is shipped with a 10-foot plug-in cord. When an existing outlet is within reach, additional electric work might not be needed.

Let me give you an idea of how much water a small pump like I use is capable of moving. Pumping water ten feet high, one of these

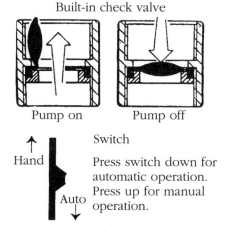

13-2
A small, but powerful sump pump that features a built-in check valve. A.Y. McDonald MFG. Co., Dubuque, Iowa

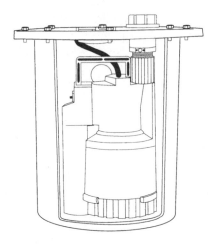

13-3
A sink-tray system. A.Y. McDonald
MFG. Co., Dubuque, Iowa

pumps can produce up to 2250 gallons per hour. That's a lot of water. The pump cuts on and off automatically with the use of a float switch. A check valve should be installed in the discharge line to prevent water that has been pumped into the vertical discharge line from returning to the sump.

A gray-water sump doesn't take up a lot of room. The sumps I use measure 15 inches tall and 15 inches wide. They weigh about 16 pounds. The supplier calls them sink-tray systems (Fig. 13-3). Call them what you want, they can provide an economical alternative to a whole-house pump station. Due to the size and design of gray-water sumps, they are not affected by small particles of debris in the same way that a single-fixture pump is. While higher priced than a single-fixture pump, a gray-water sump system is more dependable and can handle a lot more water.

Black-water sumps

Black-water sumps work on a principle similar to that of gray-water sumps. However, black-water sumps can receive the discharge of toilets and other fixtures of a similar nature. The pumps used in black-water sumps are of a different type than those used in gray-water set-ups. Black-water sumps are normally installed below a finished floor level. Quite often they are buried in concrete floors. It is possible to install such a sump in a crawl space by simply creating a stable base for it on the ground.

Black-water basins or sumps are available in different sizes (Fig. 13-4). A typical residential sump might be 30 inches deep with an 18-inch diameter at the lid. Basin packages can be purchased that include all parts necessary to set up the basin (Fig. 13-5).

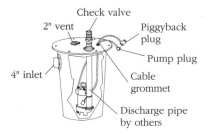

13-4
Typical black-water basin set up for use in basements or other locations. A.Y. McDonald MFG. Co., Dubuque, Iowa

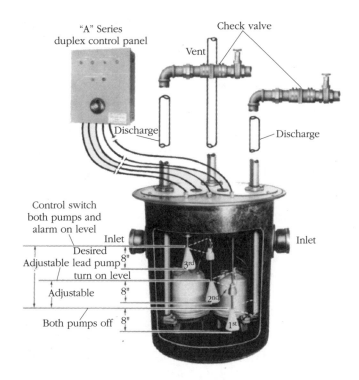

13-5 *A duplex sewage ejector system.* Goulds Pumps, Inc.

A 2-inch vent pipe should extend from the basin cover (Fig. 13-6) to open air, outside of a building. The use of a check valve is required in the vertical discharge line. A 4-inch inlet opening is molded into the basin to accept the waste of all types of residential plumbing fixtures. Some type of float system is used to activate the pump that is housed in the basin. The exact type of float system depends on the type of pump being used.

Pumps used for a typical, in-house, black-water sump are known as effluent pumps (Fig. 13-7) and sewage ejectors. The discharge pipe

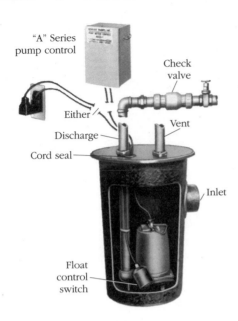

"A" Series pump control

Check valve

Either

Vent

Discharge

Cord seal

Inlet

Float control switch

13-6
Single-pump sewage ejector system. Goulds Pumps, Inc.

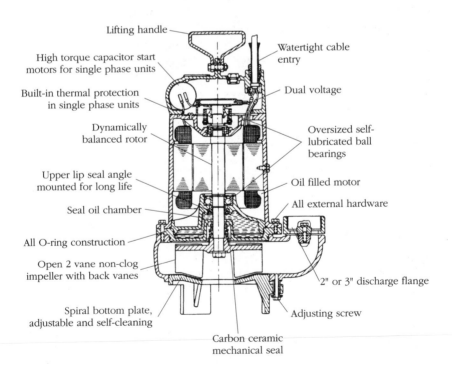

Lifting handle

High torque capacitor start motors for single phase units

Watertight cable entry

Built-in thermal protection in single phase units

Dual voltage

Dynamically balanced rotor

Oversized self-lubricated ball bearings

Upper lip seal angle mounted for long life

Oil filled motor

Seal oil chamber

All external hardware

All O-ring construction

Open 2 vane non-clog impeller with back vanes

2" or 3" discharge flange

Spiral bottom plate, adjustable and self-cleaning

Adjusting screw

Carbon ceramic mechanical seal

13-7 *Effluent pump.* A.Y. McDonald MFG. Co., Dubuque, Iowa

from the pump normally has a diameter of 2 inches, although a 3-inch discharge flange is available. Even if the waste of a toilet is being pumped, a 2-inch discharge pipe is sufficient.

The cost of a complete black-water sump system runs into several hundred dollars, but this is much less than the cost of a whole-house pump station. If you have a basement bathroom that requires pumping, this type of system is ideal. I've installed dozens, if not hundreds of them, and I've never been called back for a failure. During my plumbing career, I have responded to failures in similar systems. The most common problem is a float that has become wedged against the side of the sump. This won't happen if the system is installed properly.

If you have a house where most of the plumbing can be drained by gravity, and only a few fixtures need to be pumped, such as in the case of a basement bathroom, an in-house black-water sump is the way to go (Fig. 13-8). You can refer to performance charts, available from manufacturers, to select the right pump for your job (Fig. 13-9).

Whole-house pump stations

Whole-house pump stations are about as bad as it gets for a builder. These systems involve a lot of material, labor, and costs. Also, should the pump fail, none of the house plumbing can be used. As bad as they are, pump stations are sometimes a blessing. If you have land that cannot be built on under any other circumstances, pump stations look pretty darn good.

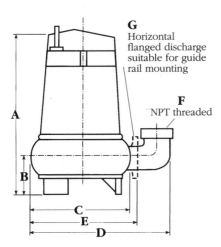

G
Horizontal flanged discharge suitable for guide rail mounting

F
NPT threaded

13-8
Dimensions and specifications for effluent pumps and sewage ejectors.
A.Y. McDonald MFG. Co., Dubuque, Iowa

*U.L. Listed, C.S.A. Approved

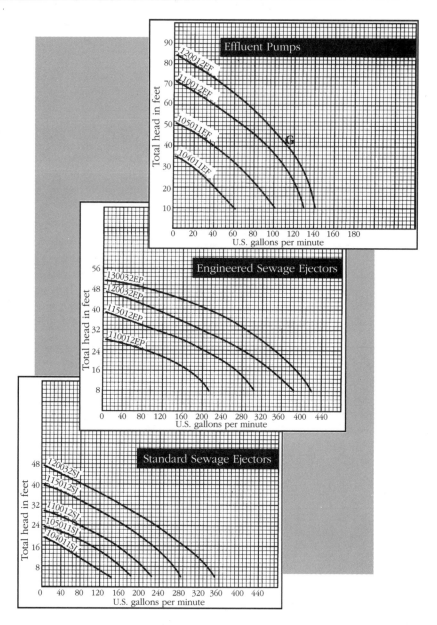

13-9 *Performance charts for effluent pumps and sewage ejectors.*
A.Y. McDonald MFG. Co., Dubuque, Iowa

Whole-house pump stations are usually, but not always, located outside the foundation walls (Fig. 13-10). Since all of the plumbing in a house with a pump station is dependent on a pump's operation, you must take some special precautions when installing such a sys-

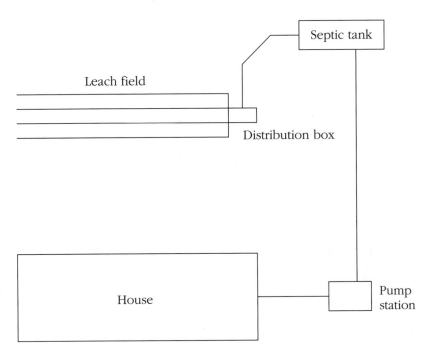

13-10 *Example of a pump-station septic system.*

tem. Code requirements are often very stringent on this issue, although they vary from place to place. We are going to talk more about code issues a little later.

It is standard procedure to install a whole-house pump station in a location outside of the house. A sewer pipe runs from the house or building to the storage sump. The size and capacity of this sump is determined by local code requirements and anticipated use. A sump can be made of many types of materials, including concrete and fiberglass.

Fiberglass sumps are not uncommon. One fiberglass sump that I know of is 3 feet deep and has a 30-inch diameter. This particular sump is designed for interior use with a grinder pump. It has a 4-inch inlet and an outlet for a 2-inch vent. The basin kit comes complete with a control panel, a hand-off automatic switch, a terminal strip, and an audible alarm. Three encapsulated mercury switches provide positions for on, off, and alarm. This is only one example of a whole-house pumping system. Many others are available.

The potential designs for a whole-house pump system are too numerous to detail. I can, however, explain how I installed one such system in a house I built for a customer.

The house that required a full-scale pump station set down low on its building lot. Even with a full basement, half of which was exposed above the ground, due to bedrock, a pump station was needed. A septic field was located behind and above the house, at a considerable distance from the foundation. Chambers had to be installed in the leach field. The basic septic system was a gravity-fed system, but the sewage from the house had to be pumped up to the septic tank. Once waste was delivered to the tank, the septic disposal process could take its normal course.

To accomplish our goal, a black-water sump was installed underground, outside of the house. I think it was about 15 or 20 feet from the foundation, in the rear of the house. Plumbing fixtures in the

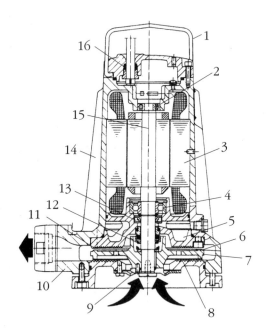

FEATURES

1. Lifting handle
2. Self-lubricating ball bearing, needs no servicing
3. Stator, insulated against heat and humidity to class F (155° C)
4. Oversized single row ball bearing
5. Oil chamber for lubrication and cooling the seal assemblies
6. Back vanes on impeller
7. Dynamically balanced impeller
8. Adjustable spiral bottom plate for handling fibrous material

9. Patented hardened rotor and stator cutter elements
10. Volute with centerline discharge suitable for mounting to guide rail bracket or discharge elbow
11. Spiral back plate
12. Mechanical lower seal enclosed in Buna N boot
13. Upper lip seal angle mounted for long life
14. Motor housing with large cooling fins
15. Rotor shaft assembly dynamically balanced
16. Watertight cable joint with strain relief

13-11 *Grinder pump.* A.Y. McDonald MFG. Co., Dubuque, Iowa

home drained by gravity to the holding tank. A grinder pump (Fig. 13-11), which was float-operated and equipped with an alarm system, pumped sewage up the hill to the septic tank. Since a fairly large holding tank was used, the pump didn't have to cycle each time a toilet was flushed. The holding tank was large enough to allow occupants of the house to use minimal plumbing fixtures for a short time, even if electrical power or a pump failed.

If for some reason the grinder pump failed, lights and an audible alarm would come on in the house. The alarm would alert the homeowners of a problem, in an attempt to prevent flooding the holding tank, also known as the pump station. This system worked well when it was installed, and it is still functioning just fine, to the best of my knowledge, after several years of use.

Requirements for pump stations vary from place to place. Even plumbing codes disagree on some points of the installation procedures. For example, one major plumbing code requires the use of two pumps in certain pump stations. This provides a back-up pump in case the main pump fails. Another major plumbing code doesn't require a second pump. For reasons like this, you must investigate the code requirements in your area to avoid costly mistakes. It's also important to consult performance charts (Fig. 13-12) to see that a pump can pump high enough to reach a disposal site.

Regardless of how they are installed, pump stations add to the cost of a project. This is something all builders should know. If an alarm system is required, and it is in many cases, the cost is going to

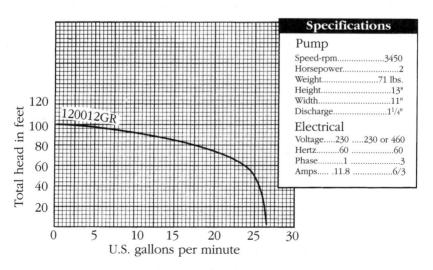

13-12 *A grinder pump sump basin specifications.* A.Y. McDonald MFG. Co., Dubuque, Iowa

be higher than if one is not mandated. Before you bid a job with a pump station, get all of your facts straight.

Code requirements

Code requirements can vary from town to town. They often change from state to state, and big differences can exist in various parts of the country. There are three major plumbing codes. One code is prominent in southern states. Another rules the western states, and a third code controls many of the eastern states extending from Virginia to the Canadian border.

There is something about plumbing codes that all builders should know. Every jurisdiction has a right to adopt and amend a plumbing code. These amendments can make the rules in one town or city different than they are in another. This can get quite confusing if your work area covers several jurisdictions. I once worked as a plumber in an area where a different license was required for every city and town. I might work in seven or eight different jurisdictions, all in the same day. Most of the code requirements were consistent, but there were enough differences to keep me on my toes.

You are probably accustomed to code changes from place to place. Building codes, like plumbing codes, tend to change from one community to another. It is, however, imperative that you not take code differences lightly. As I mentioned, one code requires two pumps in some pump stations while another code requires only one. If you were accustomed to working with the code that requires only one pump and strayed into a jurisdiction that calls for two pumps, you could be out hundreds of dollars if you assume one pump is enough. Don't take any chances. Always confirm current, local code requirements before submitting a formal bid to a customer.

I'm pretty familiar with all three of the major plumbing codes. They share many similarities, but there are substantial differences. I'd like to give you some basic guidelines for code requirements when installing pump systems. To do this, I'm going to refer to the plumbing code that I've used for most of my career. Keep in mind that the requirements in your area might very well differ from those I'm about to relate to you.

Within my local plumbing code are several requirements for pump systems and sumps. For example, a sump for a sewage pump or ejector must have a minimum diameter of 18 inches and a minimum depth of 24 inches. The code's intent is to ensure that a pump runs for at least 15 seconds once it is activated. If a larger sump is needed to accomplish the time limit, then one must be installed.

Sewage pumps and ejectors that are required to pump the discharge from toilets must be capable of handling spherical solids with a minimum diameter of 2 inches. Gray-water pumps must be able to handle spherical solids with a minimum diameter of ½ inch. A check valve must be installed on the discharge line of any type of pump in a sump.

When sizing a sump, the basin must not be capable of holding more than one half of a day's expected discharge. In other words, if a daily discharge rate were set at 200 gallons, the sump could not have a capacity of more than 100 gallons. A smaller capacity is always acceptable, as long as the pump runs for at least 15 seconds in every cycle. Covers for sumps must be sealed to be gas-tight.

If a pneumatic ejector is used to pump waste from a sump, a special relief vent is required to instantaneously relieve the pressure in the receiver.

Duplex pumping (the use of two pumps) is allowed under my local code requirements, but not mandated. This, as I've already said, is not the case with all plumbing codes. The minimum capacity of a sewage pump or ejector with a 2-inch discharge pipe, according to my local code, is 21 gallons per minute.

I've covered the key points of code requirements in my area for pump stations. Some plumbing codes are more extensive than others, so don't take these examples as law. Check with your local code enforcement office before you make any solid plans or offer firm quotes on jobs.

Would you like to save some money on the cost of your next job? Our next chapter shows you how to minimize the cost of well systems, and Chapter 15 does the same thing for septic systems. Let's move on.

14

Keeping costs down on well systems

How can you keep your costs down on well systems? There are many ways to reduce or limit what you pay for a well. Some of them have to do with field conditions. Others depend on your ability as a negotiator. Is it really possible to keep well expenses in check? Sure it is. You need to exert a little effort, but the savings can be considerable.

In this chapter, we are going to look at a number of ways to reduce or hold your costs on well systems. This includes not only wells, but the pump systems that are used with them. I plan to give you a number of examples, so that you can see my suggestions in action.

At this point, you have learned a great deal about wells and pump systems. The foundation that's already laid gives us solid ground to work with in this chapter. Since we are going to discuss cost overruns in Chapter 18, I am not going into great detail on that subject as we progress through this chapter. However, I am going to show you how to make the most of what you have already learned.

Cutting too many corners

Cutting too many corners to save money is dangerous. You can damage your reputation, offend customers, and suffer other consequences. I don't recommend providing substandard materials or workmanship on any job at any price. Even if a customer asks you to provide low-quality work, I suggest that you not do it. If you do intend to do it, get all of the facts of your agreement with the customer in writing. Let's look at some examples that illustrate the risks of working below normal standards.

Using a spring

Using a spring as a primary water source can put you at risk in many ways. Springs are a cheap source of water, but the water might not remain available throughout the year. It's possible that the water source could dry up in the summer or freeze in the winter. As a builder of a new house, you have a warranty period to honor. If the spring you use for a water supply fails to produce water, you might have to provide the customer with a well at no charge. Paying for a well and a pump system out of your own pocket is not a pleasant experience.

When you bid a house that requires a private water source, you should normally plan on a well system. You need to determine what type of well system should be installed. We are going to talk more about this a little later. But, what should you do if a customer insists on having a spring as a water source?

Let's say that you have a couple who wants you to build them a custom home in the country. The house needs a private water supply. It so happens that the land where the house is to be built has a good spring. In fact, the person who sold this couple the land told them that the spring had been used for drinking water for years. Based on representations made to them by others, the couple feels the spring would be adequate for their household water needs. They have instructed you to bid the job based on using the spring as the only domestic water source. What are you going to do?

If you have enough experience, you might counsel the couple about the risks associated with using a spring as the only water source. Should you not have an adequate grasp of facts pertaining to the use of various water sources, it would be appropriate to suggest that the couple talk to experts before they commit to using the spring.

You could at least explain the risk of having the spring run dry. Then you might mention, depending upon geographic locations, the risk of having the spring freeze during winter months. Contamination of the water is another issue that you might bring up. The cost of enclosing the spring and making it acceptable as a source of potable water should definitely be discussed. Many people have misconceptions about the use of a spring as a domestic water source.

I would guess that a number of people would tell you that all that is needed to use a spring is to put a pipe in it and pump the water to the house. You now know this is not true. Some type of containment should be placed around a spring that is going to be used as a water source. It's very important to protect the spring from any entry of groundwater. This work costs money. Granted, it doesn't cost nearly as much as digging or drilling a well, but it is an expense that some people don't anticipate.

In addition to the cost of a casing or lining for the spring, trenching and fencing might be needed. Springs located on the side of a hill should have trenched-in drain pipe installed on the uphill side of them. These slotted drain pipes collect surface water running down the hill and divert it away from the spring. Fencing might be needed to keep livestock from contaminating the spring.

Once you, or some other expert, explain all the facts about springs to your customers, they might need some time to think over their options. If the couple still wants to use the spring, you might have to succumb to their wishes if you want to build a house for them. But, don't do this without some protection.

Assuming that your customers insist on using the spring, you should ask them to put it in writing. This written agreement is your protection. If the couple puts their orders to you in writing and they assume all responsibility for the quantity and quality of water provided by the spring, you can limit your exposure to only the pump system. This approach should be acceptable to any reasonable person. If the couple is unwilling to sign an agreement, there might be a good reason to avoid doing business with them. I hate to say it, but sometimes people set other people up in order to take advantage of them. My personal experience has proved this to be true far too often.

What might happen if you don't get a liability waiver from the couple in our example? Maybe nothing, but a lot could happen. If you don't have anything in writing, the circumstances surrounding your work might be presented very differently in litigation. For example, the couple might tell a judge during a legal dispute that you recommended the spring, without ever advising them of the possibilities that might make the spring unsuitable.

If a judge were given a false impression of what actually went on between you and your customers, you might be found responsible for inadequate or unsuitable drinking water. As an experienced builder, you might be judged to be much more knowledgeable than your customers on the subject of private water supplies. All in all, you might be ordered by the court to provide a conventional, suitable water source for the customer at your own expense. If this were the case, the cost you incur could easily run in excess of $4000. A simple little liability waiver could save you this aggravation and expense.

Cheap pumps

One way to reduce the cost of a well system is to install cheap pumps. They might, however, be more trouble than they are worth. Some builders feel that the pumps installed for new houses don't have a direct effect on their businesses. I disagree with this kind of thinking.

Most builders don't install their own pump systems. They typically subcontract the work out to specialized contractors. In doing this, the builders put a buffer between themselves and the work that's done by the subcontractors. Materials, like pumps, might be provided by builders or by the subcontractors. My experience has shown that most installers prefer to provide the materials for their jobs. When this is the case, a pump failure ultimately rests on the shoulders of the installer who supplied the pump.

A builder who hires a pump contractor to supply and install a pump doesn't have to assume any cost to have the pump repaired or replaced while it is under warranty. The pump installer has to make good on the work done. This doesn't mean, however, that the customer won't blame the builder for the pump trouble. When a pump fails, people tend to get upset. If you want to avoid disgruntled customers, you should use the best materials available that fit into your budget.

There is not a lot of cost difference between a good pump and a questionable pump. The difference in cost between a nylon insert fitting and a brass insert fitting is minimal compared to the added benefits provided by the brass fittings. Using only one stainless-steel clamp per pipe joint saves a few dollars on a typical well installation, but the savings is not worth the risk, in my opinion. If you cut every material corner that you could, the savings might reach a couple hundred dollars. This amount of money is nothing to be sneezed at, but why would you risk your reputation, your customer satisfaction, and the cost of call-backs for such a small savings? I just wouldn't do it. I'm all for saving money and reducing job costs, but I won't do it at the expense of my customers or my reputation.

Smart ways

There are many smart ways to reduce the costs of wells and well systems. Many of the options don't have any effect on the quality of work or materials you provide your customers. These types of savings are well worth considering. Let's do that right now.

Well selection

Well selection can have a lot to do with the cost of a job. You now know that driven wells are your least expensive well option. Dug or bored wells are less expensive than drilled wells. Knowing this, you have some room to work on your budget, subject to local conditions.

When you do a site inspection, you have an opportunity to choose the best location for the well. There is more to choosing a well location than just finding a spot where the well is out of the way. For example, when a well is placed 50 feet farther from a house than it needs to be, it increases the cost of a job because extra trenching and pipe is required.

The added length could also require you to upgrade the pump. If a well is at a distance and depth that is borderline between a half-horsepower pump and a three-quarter-horsepower pump, the extra 50 feet of distance would clearly call for the larger pump, increasing the cost of the pump system. The combined cost of trenching, pipe, and increased pump size could amount to hundreds of dollars. Assuming that there is no good reason for extending the distance to the well, keeping it closer to the home provides a cost savings without sacrifice to you or your customer.

Shallow wells

Shallow wells are much less expensive than drilled wells. If you have an option between these two types of wells and feel that a shallow well is adequate, there is no reason to spend extra money for a drilled well. While drilled wells do offer advantages over shallow wells, the additional cost might not be justified by the differences.

Depending on ground conditions, you might not have any choice in the type of well you choose for a job. Your customer might spec out a drilled well, regardless of geological conditions. If this happens, you have little choice in the matter. It might be prudent to make your customer aware of the possible cost savings by using a shallow well, but I'd steer clear of recommending one.

If your customers specify a drilled well and you talk them into a shallow well, you could wind up in a world of trouble down the road. Because you talked them into a shallow well to save them money, you would bear responsibility, in my opinion, if the shallow well did not perform to the customer's satisfaction. You are usually safe when you recommend a superior product or service, but you can get into deep water fast by recommending something that is of a lesser quality.

Driven wells

Driven wells are the least expensive well that can be created. Sometimes these wells are adequate for a full-time residence. But, there are also times when driven wells can't keep up with the water demands. Normally, driven wells should not be installed for new houses.

If you are going to install a driven well for a customer, I feel that you should treat it in much the same way as you would a spring, which we discussed earlier. Driven wells, in my opinion, should only be used for special circumstances. Trying to save money with the use of a driven well can be very risky.

Pumps

Pumps are a necessary part of well systems. As a builder, you might select the type of pump to be used for a new home. Customers often tell you what type of pumps they want. Sometimes the decision is made by the pump installer. If you or your installer choose a pump for a customer, you are on the hook if something goes wrong. The only time you don't have to bear the burden of responsibility for pump performance (outside of an improper installation) is when a customer instructs you to use a specific pump. This means that you owe it to yourself to become aware of pump selection and applications.

As you know, the three pumps available to you for routine residential installations are single-pipe jet pumps, two-pipe pumps, and submersible pumps. I've stated that submersible pumps are my preference for deep wells. Can you save some money by installing a two-pipe pump for a deep well? It's very possible that you can. Should you take this approach? It really depends on your customers and your circumstances. Typically, I would recommend using a submersible pump, but there are some disadvantages to a submersible pump.

Disadvantages of submersible pumps

The list of submersible pump disadvantages is not very long. One possible disadvantage of a submersible pump is the need to pull the pump out of a well if it fails. When a two-pipe pump fails, it is normally very accessible for repairs. This accessibility is comforting to some people, and it can result in lower repair costs.

I've heard that lightning is more likely to strike a submersible pump than a pump that's installed within a home. Whether or not this is true, I don't know. It has been my experience that more submersible pumps are affected by lightning than jet pumps. In fact, I don't know of a single instance when a jet pump has been hit by lightning. But, I do have personal knowledge of several submersible pumps that were struck by lightning. This might not seem like a big deal, but if a pump is not covered by lightning insurance, the replacement cost can run several hundred dollars. Admittedly, lightning strikes might not be a good reason for choosing an above-ground pump, but it is a potential consideration.

Cost is another disadvantage of submersible pumps. In my opinion, submersible pumps are a good value, but they tend to be more expensive than above-ground pumps. Money is always some part of a decision-making process for builders. Even if a submersible is worth every penny of its cost, such a pump might not be needed.

Two-pipe pumps

Two-pipe pumps can be used for deep wells. These pumps require two pipes, and that is a disadvantage in two ways. The cost of a second pipe requires the overall price of a pump system to increase. Perhaps more importantly, the need for a second pipe creates twice the risk of pipe failure. Two-pipe pumps have pros and cons.

Submersible pumps can move more water more efficiently than two-pipe pumps. This advantage might not matter to a homeowner. Having an above-ground pump sitting in a closet produces some risk to a home. If something goes wrong with the pump system, the home could be flooded. Noise is another factor to consider when deciding between a submersible pump and an above-ground pump. Pumps installed in basements or the common living space of a home can be noisy. Submersible pumps cannot be heard when they run.

I've had homeowners complain about being able to hear water running down a drain located in some wall of their homes. If people complain about the relatively quiet draining of water, they are certainly likely to complain about the running noise of an inside pump. If you are thinking about installing a pump system close to living space to reduce the cost of a water distribution system, you should also carefully consider the noise factor.

Single-pipe pumps

Single-pipe pumps have very limited uses. If water has to be lifted more than about 25 feet, single-pipe jet pumps become questionable in their ability to perform. This is not to say that a jet pump can't pull water up to 30 feet. It is also possible that a jet pump won't be able to raise water more than 20 feet. The elevation above sea level is a factor in how well a jet pump, which works on a vacuum basis, can perform.

Installing a single-pipe pump and expecting it to lift water more than 25 feet is a risk. While you can save some money with a single-pipe pump, you must not install this type of pump under questionable circumstances. Common sense goes a long way in day-to-day decisions.

Circumstances often dictate whether or not a single-pipe pump can be used. Just as you can't drive a well point through bedrock, you can't expect a simple jet pump to pull water up from a 100-foot-deep well. The main question you normally have to deal with is a decision between jet pumps and submersible pumps.

Pressure tanks

Pressure tanks should be used on all residential well systems. The size of a tank is relevant to the pump that you are using. You already know that larger pressure tanks reduce the running time of a pump over a period of years. However, a super large tank is not usually necessary. As long as the tank being used is sized in conjunction with the pump, it is not necessary to oversize the tank.

Never try to save money by eliminating the use of a pressure tank. I'm quite sure you would regret it if you did. Pressure tanks that hold between 20 to 40 gallons generally work out very well. Larger tanks don't hurt anything, but their additional cost might not be worthwhile. Before you can justify an extra-large tank, you must have some special considerations to take into account.

Shopping prices

Shopping prices can really pay off. Some suppliers charge more than others. This should come as no surprise. Competitors in most fields charge varying prices. It stands to reason that some prices are going to be higher than others for the identical product.

Being a builder, you might not buy your well pumps and associated equipment directly. You might buy it indirectly, through your pump installer. Either way, shopping pays off. You can shop individual material prices or you can shop subcontractor prices.

It doesn't hurt you or your customers to look for lower prices. However, it's important to keep a good perspective. Going with the low bidder on a labor-and-material bid can be very risky. This is not the case when you are looking only at material prices. If you get three prices on an identical pump, you should have nothing to lose by going with the lowest price. It is not until you factor in the quality of workmanship, dependability, and similar human traits that it becomes difficult to rely on prices.

Experience can go a long way towards helping you choose subcontractors. Until you have a lengthy track record with subcontractors, you don't know what to expect. Regardless of what is promised, you can't be sure of what you are going to get. However, after doing several jobs with the same subs, you begin to get comfortable with them. This in itself can be dangerous. If you get lulled into a sense of security, you can come up short at some point.

Most installers prefer to provide their own materials. There are a couple of reasons for this. First, when installers supply the materials, they don't have to depend upon you to put the proper materials on a job. This saves time, but it can cost you money. Second, most sub-contractors who supply their own materials mark up the costs. This isn't entirely bad if you look at all sides of the deal.

If you assume that the mark-up on a well system is $100, it might very well be worthwhile to you to let the pump installer supply all materials. A mark-up of $300 is a little harder to swallow. Since there are no set rules on how much mark-up can be added to the cost of products, you have to educate yourself. If you don't, there is no way for you to know how much you are being charged for materials.

Some material suppliers won't sell pumps and related materials to builders for the same prices that they charge plumbers and pump in-stallers. This might not be fair, but it does happen. If you check ma-terial prices, you might find that your pump installer can have a strong mark-up on the products without having much effect on your costs. Let me give you an example.

Assume that you need a complete pump set-up for a deep well. You are going to be installing a submersible pump and a stand-type pressure tank. For the sake of this example, assume that the whole-sale cost for your plumber is $750. Further assume that the plumber adds a hefty mark-up to the wholesale price and gives you a figure of $900. Is this too much for you to pay? It depends.

Let's say that you can't buy direct from plumbing wholesalers. This is basically the case in Maine. Plumbing wholesalers in my area sell only to licensed plumbers. A builder can't buy from them, at least in theory. I know of a few nonlicensed people who deal with these suppliers, but most can't.

If you have to go to a building supply center or hardware store to buy a pump, you might get a 10 percent discount from the retail price. Even with your discount, the price might be more than what your plumber is charging you. I know this can be true, because I've seen it from both sides of the table. As both a licensed builder and master plumber, I have better buying power than most builders. Sometimes I can buy something through my plumbing company at wholesale price, mark it up 30 percent for my building company, and still pay less than what it would cost at many stores with my builder's discount. In my personal circumstances, this doesn't mean much on a paper trade. But, it could mean a lot to you.

Think about what I've just told you. It might be just as cost-effec-tive for you to pay your plumber an inflated price as it would for you to pay your supplier a discounted price. This makes everyone happy.

Your plumber makes some extra money, and you don't lose any money. You might even save a few bucks. This is something you have to weigh out carefully.

Read the specs

When you are pricing materials, read the specs carefully. You should also pay close attention to the proposals that are supplied by the installers. A few little gimmicks in a proposal can make a big difference in the bid price.

Two well pumps that appear identical could be very different. Motor size is one factor to consider, but so is the production rate of the pump. Many good brands of pumps are available. The brand you choose is likely to depend on the dealers in your area. Although I don't have a particular favorite, I have a lot of confidence in two name brands. If I'm calling the shots, one of these two brands is used.

As a plumber, I've worked on many different brands of pumps. It's easier to get parts for some than others. It seems like some brands are more prone to failure than others, but this might not be true. Just because my plumbers repair a lot of pumps from one particular manufacturer doesn't necessarily mean that the manufacturer produces bad pumps. It could simply be that the brand is more popular than other brands. If an area has 25 brand-A pumps, 2 brand-B pumps and 5 brand-X pumps, it stands to reason that there should be more failures reported with brand-A pumps. It's purely a numbers game.

I believe that some pumps are better than others. Without question, some pumps are easier to work on than others. Each manufacturer usually has something special to offer as a feature or benefit. You can save some money by purchasing no-name pumps, and they might do a fine job. However, you are normally better off sticking with major brands that are used regularly in your area. If nothing else, it is going to be easier to find parts.

Most architects specify materials very clearly. Homeowners rarely do. Depending upon who is providing you with job specifications, you might know exactly what materials to bid, or you might have no idea of what is wanted. If you have questions, ask them. The type and brand of pump you figure into your bid could decide who wins the job. Don't take chances on this. Get a clear understanding of what is wanted, and bid the job accordingly.

Watch the proposals given to you by subcontractors for little phrases that could come back to haunt you. Let's say, for example, that you spec out a brand-D submersible pump for the installer to bid. If two contractors follow your instructions and bid the work with a brand-D pump and one contractor bids the job with a brand-X pump, the bid with the different pump won't be comparable to the others. This tactic is used from time to time by contractors striving to be the low bidder. Some of these contractors are very good at hiding their true intentions. You might never know that the pump they used wasn't the type you specified.

A tricky contractor can use a common phrase to fool you. Being a contractor, I assume you are familiar with the old, "or equal" clause. Who's to say what is equal? When I get a quote or proposal with an "or equal" clause, I strike through it. If contractors can't commit to using the materials I specify, they don't work for me.

How many builders hang around and watch an entire pump installation? Not many. Do you know how easy it would be for me to switch pump types on you? If you're like most builders, it would be very easy. I could even substitute a used pump without you knowing. These types of switches are very easy when submersible pumps are installed.

Once a pump goes down into a well, it doesn't normally surface again until something is wrong with its performance. So, if I stick a cheap pump in your well and charge you for an expensive pump, I've made a few extra dollars. If I'm really tricky, I have one of the good pumps set out in the open when I arrive on the job and as I'm working. At the last moment, I switch to a used or cheap pump. If you happen to stop by to look over the materials or to chat, you see the good pump or maybe just a box for a good pump. What you don't see is the inexpensive pump I've installed in place of the pump you think you are getting.

I'm not saying that pumps are switched on the job very often, but it can happen. And really, who's going to know the difference until the pump fails? By then, the warranty has probably expired and the rip-off contractor might be long gone. I don't want to scare you, but it seems to me that you should be aware of what could happen on an unsupervised job.

Do your own installations

If allowed by local authorities, it might be worth it to you to do your own installations. The installation of a pump system is not very complicated. If you have basic mechanical skills and can follow instructions, you can install a pump and pipe up the accessories. Doing this work yourself can save you hundreds of dollars.

Some areas allow individuals who are not working under the supervision of a master plumber to install pump systems. The regulations generally require the installer to stop after the installation of the pressure tank. In other words, you might be allowed to install the pump and related accessories, but not the water distribution pipes of a house.

If you can get a limited license or if no license is required, there is an opportunity for you to pick up extra profit from each job. In my area, the labor to install a submersible pump system, not counting the cost of trenching, is usually around $750. Since the work can be done in about half a day by an experienced installer, this isn't a bad rate of pay. I don't know if you would be allowed to do your installations, but it's worth considering.

Negotiations

One option for getting lower prices is to negotiate with your well and pump installers. When a bid price is submitted to a general contractor from a subcontractor, there is often some room in it for haggling. Simply asking for a lower price might be all that's required to get one. If your subcontractors have figured the job tight and won't budge on their numbers, there is another card you can play.

Wells don't generally have to be installed at any particular time or phase of construction. Since most contractors hit dead days, you might be able to use these broken days to your advantage. Subcontractors might be willing to give you a lower price if you don't pin them down to an exact installation date. If a sub knows that your job is ready and waiting when a dead day comes along, the sub can work on your job. This is much better than sitting in the shop, waiting for the phone to ring. Since your job is being treated as fill-in work, you might see a lower price.

I've often used the fill-in procedure to lower my costs with subcontractors. Rarely have I encountered trouble from this approach. However, make sure that a time limit is set. This limit should be explained in your written agreement. For example, you might note that the well is to be drilled within 45 days from the date of your agreement. This protects you and still gives the well driller a wide latitude within which to work. Creative ideas like this can definitely work to your advantage.

Volume deals

Volume deals can pay off for you by the end of a year. If you are in a position to offer your well and pump installers several jobs in close proximity of each other, you should be able to get a discount.

When I was building at peak volume, I was producing about 60 houses a year. It was common for me to start 10 houses in 10 days. Knowing that I had these houses coming up, I was able to negotiate some very good volume discounts with all of my subs. All of the plumbing for the houses was done by my own piece workers, so discounts weren't a factor in the pump installations. The well work, however, was done by subcontractors, and I was able to get price breaks by giving them several contracts at the same time. By offering multiple contracts and a flexible production schedule, my well installers gave me great prices. Not all builders are in a position to offer multiple contracts, but if you are, it should be worth a discount.

Now that we have covered the issue of keeping costs down on well systems, let's take a similar approach to septic systems. Turn the page and let's get started.

15

Making a habit of installing low-cost septic systems

It doesn't mean that you are cheating your customers when you make a habit of installing low-cost systems. Actually, you might be doing them a big favor by striving for a low-cost septic system. If you pass the savings along to the consumer, you have used your knowledge to save them money. What customer wouldn't appreciate this type of surprise?

Builders can face some very tough times during their careers. I know I have. Surviving economic downturns and recessions are just some of the obstacles to overcome. Increased competition is always a threat, and there are many other factors that can put a builder out of business. By honing your money-saving skills, you can improve your odds of survival.

Making money is very nice, but it requires you to pay taxes. Saving money can be even better than making money, because the tax bite is not so vicious. I've yet to meet anyone who enjoys paying taxes. Most of us do it, but few of us feel we get our money's worth.

The cost of living goes up almost constantly. This means that our incomes should rise with the increases in the cost of living. Did your income rise the year before last? How much more money did you make last year? Are you on track to make more money this year? A lot of businesses, including building businesses, suffer from declining sales and income. Left alone, this pattern can eventually drive a business out of existence. If you are feeling a pinch in your pocketbook, you must look for new ways to make or save money.

One way for rural builders to reduce the cost of doing business comes in conjunction with septic systems. Some systems simply cost less than others to install. We've talked about this issue already. But, you must get involved if you want to be assured of reduced septic costs time and time again. Spec builders have to look for good land. Custom builders sometimes have to look for alternatives to high-priced septic systems. If you can get into the swing of doing this right, you can make and save more money.

Knowledge

Knowledge is one of the best tools that a builder can possess. If you know enough about what you're doing, you should be able to do it well. More knowledge in some work area, such as septic systems, can give you a competitive edge. You've already taken a giant step in the right direction by reading this book. The fact that you are willing to invest valuable time reading about one or two phases of your building business proves that you care about your business and your customers.

Informed builders are better builders. This book is an excellent starting point for you, but it is not the only source of knowledge that you need in order to become more competitive. Look for additional reading materials from local and government agencies. Agencies such as the United States Environmental Protection Agency can provide you with valuable information. Local and state agencies might also be able to offer some guidance on septic system issues.

Your subcontractors are a natural source of information. If you have questions about your next septic job, who is better qualified to answer them than your septic engineer or installer? Ask questions. Become informed. Get on the right track, and you can see substantial savings over the coming years. These savings can help offset any economic drops in business production.

Try installing one yourself

If local regulations allow you to install a septic system, try installing one yourself. Even if you don't make much money from the work, the hands-on experience can do a lot for you. Once you have put your feet in the trenches and your hands on the pipes, you are going to have a better understanding of how septic systems are installed. Reading about procedures is fine, but carrying out the procedures is the best way to understand them. You might even discover that installing your own septic systems is an excellent way to boost your income.

Custom builders are at the mercy of the market. So are spec builders, for that matter. When a slow economy keeps new construction to a minimum, it's important to make the most from every job. You might accomplish this by doing your own septic installations. If this line of action doesn't suit you, there is still room to make and save more money by using your head.

Choose sites selectively

Whenever possible, choose your building sites selectively. Site selection is critical to a low-cost septic system. If you can avoid chamber systems and pump systems, you're money ahead. Earlier chapters in this book provided you with a host of suggestions for finding the best septic sites. Go back and read those chapters again. Highlight key points and refer to them when you are out in the field searching for land.

Building lots and land can be in short supply. Sometimes builders have to take lots that are not as desirable as they would like. Any builder who has been in the business for a few years knows this. But, some sites simply aren't worth the price. This doesn't necessarily mean that you shouldn't buy the lots. What it does mean is that you should negotiate for a lower price.

If you have a solid understanding of soil types, you can use perk tests to drive down the cost of some building lots. Sellers are reluctant to drop prices for no apparent reason, but if you build a strong case, you might be able to influence them to reconsider their thinking. I've done this many times and under varied circumstances. This approach is very effective, so let's spend a few minutes going over the details.

A spec builder

A spec builder has a wealth of options when choosing building sites. Since spec houses can be built almost anywhere, a builder doesn't have to accept the first building lot to come along. Unlike custom builders, who must build on the land that their customers own or want to buy, spec builders have a free hand to wheel and deal. They can look for the best land bargains available. As long as the building site shows good promise for resale, a spec builder doesn't have to put emotion in front of financial logic.

Let's assume that you are a spec builder. You have found three building sites. All of them appeal to you. The three building lots are similar in size and price. Location is not a problem. A spec house built on any one of them should sell well. You only want one of the lots, so you must decide which one to buy. All three lots require a septic system and well. When all factors are considered, the lots seem to be equal in potential for a quick sale once a house is built. Which one are you going to buy?

In this scenario, there might not be much haggling to be done. Since all of the lots are about the same, you don't have a lot of room to work on the price. In this case, you might just pick the lot you like best and go with it. Or, you could try a price-lowering strategy. How can you do this? You can start by requesting septic designs for the various lots.

Once you have septic designs to review, you might find some differences in the land. One lot might require only a pipe-and-gravel septic system while the other two call for a chamber system. If this were the case, you should choose the lot that is cleared for an inexpensive, pipe-and-gravel system. With all other factors being equal, it would be foolish to buy a lot that required a more expensive septic system.

Now, let's change some of the criteria in our example. Assume that one of the three lots has a much better location than the other two. This one lot is the site you really want. But, there is a problem. The good lot requires a chamber system for its septic disposal. The other two lots require pipe-and-gravel systems. After doing some estimates, you discover that the better lot requires about $5000 in additional site work, due to the chamber system. This, in effect, puts the price of the better lot $5000 higher than the other two, even though all three are priced the same. As much as you like the one lot, your budget numbers show that the area can't support a higher-priced house. This means that you must take the $5000 out of your building profit. You don't like this idea, so you try a negotiation tactic.

After compiling cost estimates from three septic installers, you make the landowner an offer to buy the building site at a reduced cost. You are able to justify the difference because the lot requires an expensive septic system. The seller might not accept your offer, but there is a fair chance that you can get the land for a lower price. Since you are able to document and show the landowner a viable reason for making a low offer, you are in a better position to win the negotiations.

I have often been able to acquire land at reduced prices after showing sellers my reasons for making a low offer. If you can get the price of a building lot dropped to compensate for an expensive septic system, you have balanced the scales. It's true that you must still install a costly septic system, but you are compensated by the lower price of the land.

Cost overruns

Cost overruns are not uncommon among builders. When septic systems are involved, the chances of going over budget are increased. If you are building a house where the private sewer connects to a public sewer, there are fewer variables than you have with septic systems. Problems can come up with any type of work. Septic systems, however, seem to be especially prone to cost increases. There is no good excuse for this. Bidding and installing a septic system is not a difficult task. It is far more difficult to anticipate the shift in lumber prices than it is to calculate the cost of installing a septic system.

Why do so many builders find themselves in financial embarrassments when dealing with septic systems? The common denominator seems to be negligence. Too many builders make assumptions and then find out that their guesses were incorrect. I've seen this happen over and over again. And yes, it has happened to me, too.

How many times have you given customers a ballpark price? When was the last time that you took an educated guess at the cost of an electrical rough-in? Builders frequently plug numbers into their estimates based on experience. They also use square-foot pricing formulas. These approaches can be surprisingly accurate in many cases, but they should not be applied to septic installations.

Septic installations are like snowflakes, no two are identical. Many septic systems are similar enough to allow the use of an average cost when figuring a job. But, for every nine that come out on the average, one is going to cost much more. This is only an example. I don't know what the true percentages work out to be. But, I do know that if you bid low and have to pay high for a septic system, it could cost you thousands of dollars.

Get septic installers to give you firm price quotes before you make any commitments to customers. Don't accept estimates, demand quotes. A lot of contractors fall into trouble when they rely on price estimates. An estimate is not the same as a quote. If you have three solid quotes from reputable septic installers, there is no reason why you should suffer from cost overruns. The septic installer might lose money from a mistake in judgment, but you won't. Work only with firm, written quotes and you should not have to worry about coming in over budget on your next septic system.

Shop around

Shop around before you accept the bid of a septic installer. You might be surprised at how much difference there can be in the prices offered by various contractors. I'm sure you already have a good idea of the potential price spreads. Any experienced builder who has worked with subcontractors has seen wide-ranging prices for identical work.

I've gotten bids for jobs that had prices so far apart that I assumed a mistake had been made. In running down these bids, I've rarely found the great difference in price to be the result of a mistake. Some contractors simply ask for more money than others. This has proved to be true in all the trades that I've subcontracted over the years.

The septic installer I now use is a great guy. He does fantastic work for fair prices. However, when I was searching for the perfect septic installer, I got some wild bids. There have been times when one contractor has offered a bid price that was nearly double that of competitors. This is the exception rather than the rule, but vast differences do exist. It has not been unusual for me to see prices as much as 40 percent higher for identical work specifications.

As a bidding contractor, I know that it is unusual for contractors to arrive at identical prices for work. But, I consider any price that is more than 10 percent apart from a competitor's to be suspect. It might be too low or too high. Either way, I become concerned. I like to check my subcontractors out thoroughly before I give them any work. By doing this, I've often found a number of reasons for not using particular contractors. I've had the feeling that some were trying to take advantage of me. Sometimes I've sensed that a contractor didn't seem to want my work. And, I've seen a lot of sloppy estimating. In my opinion, a contractor who cannot estimate a job properly is a risk as a subcontractor.

If you don't already have a septic installer that you know and trust, find one. Solicit bids from several installers. I recommend getting at least five bids. The more prices you get, the easier it is to predict who is on target. I'm not going to take up your time telling you how to deal with subcontractors, but you should be shopping your septic prices if you want to consistently produce profitable work. Even if you have an installer who is nearly perfect, like mine, you should solicit outside bids from time to time to make sure that you are not paying too much. I'm willing to pay a little extra for the kind of service I get from my installer, but there has to be a limit as to how much extra cost is justified.

16

Common problems with well installations

Well installation problems can make life difficult for builders. When a person buys a new house from a builder and the well fails, the builder is likely to get a call. Some builders have no idea of how to troubleshoot well and pump problems, but they are the first people to be contacted. Homeowners who have wells that are not functioning properly are usually distressed. If ignored, they can quickly become grumpy.

Problems with wells and well-pump systems are not unusual. Is there anything you can do to eliminate these problems? I doubt it, but you can reduce the occurrence of such problems with quality workmanship and supervision. However, I believe you must be prepared to deal with a variety of problems associated with wells and their pumping systems.

Assuming that you use subcontractors to create your wells and to install your pump systems, you can call those people to help solve your problems. Having this option doesn't release you from responsibility, but it does make your job easier. Still, homeowners with well problems are going to be looking to you for support and help. The more you know about troubleshooting wells and septic systems, the more valuable you become to your customers. This is important.

As I'm sure you have gathered, this chapter is dedicated to dealing with problems. I wouldn't expect you to throw a tool box in the back of your truck and rush out to fix well problems. If you have enough background on well problems, you might be able to give customers helpful advice by telephone. Doing this might get their systems back in action right away. Your subcontractors are going to appreciate not having to make call-backs, and your customers are going to be pleased to have their water problems solved so quickly.

The information we are covering in this chapter is somewhat technical at times. Other aspects of the text are not so technical. I intend to break the chapter down into three primary categories. We are going to talk about problems that are indicative of various water sources. Problems with pumps and related equipment are also discussed. Another area of interest keys in on water quality. For example, if someone calls you and complains about a nasty smell in their water, you can refer to this chapter for advice on what might be causing the odor. Just for the record, sulfur would be the most likely cause of this problem. Let's start with basic well problems.

Basic well problems

Basic well problems are not common. In this category, we are talking about trouble with wells themselves, more so than with pumps. Water quality is not a part of our present discussion. Since well problems are often segregated by the type of well that's used, we are going to investigate the problems on the basis of specific wells.

Driven wells

Driven wells don't usually have a large holding capacity. It is not uncommon for driven wells to have low flow rates or low recovery rates. Both of these factors can contribute to a well running out of water. If a customer with a driven well calls to complain about having no water, you might be faced with a pump problem or a well problem. This is true of all types of wells. Driven wells are the most likely type of well to run dry. This is a simple problem to troubleshoot.

If you suspect that a driven well is out of water, you can gain access to the well and drop a weighted line into the well to determine if any water is standing in the well. If there is little or no water present, your problem might be with the point filter or the water source. If the well point has become clogged, water can't enter the well pipe. Assuming that there is insufficient water, you can take one of two actions.

If your area is experiencing an extremely dry spell, you might have to pull the well point to identify the problem. This is not an easy task. However, if area conditions don't point to a drop in local water tables, you might ask the customer to avoid using any water for a few hours and then try the pump again. Given several hours for recovery, the well might produce a new supply of water. This doesn't rule out a partially clogged point, but it tends to indicate a low flow rate. To be sure of what is going on, the point must be pulled and inspected.

Due to the nature of a driven point, your options for finding out if water is present in the water table are limited. If water can't pass through the filter of a point, water won't enter the well. This is not the case with other types of wells. Unfortunately, some of the money saved by installing a well point can be lost through later problems, such as having to pull the point for inspection and possible replacement.

Sand in a water distribution system that is served by a well point is an indication that the screen filter on the well point has openings that are too large. This type of problem can be addressed with the addition of an in-line sediment filter, but the true solution lies in replacing the well point with one that has a finer screen.

Other contaminants can enter a water distribution system through a well point. These entries into the water system can be filtered out with water treatment conditioning equipment. Replacing a well point might help solve this problem. Essentially, some type of conditioning equipment is probably needed to eliminate very small contaminants.

Shallow wells

Shallow wells are less likely to pose problems than driven wells. However, these wells can provide builders with head-scratching trouble. Some shallow wells cave in over time. It doesn't always take a long time for this to happen. A new well can experience problems with cave-ins long before warranty periods are over. This is not a common problem, but it is one that could happen.

Shallow wells sometimes run out of water. Given some time, these wells normally recover a water supply. If a shallow well runs dry, there is very little that can be done. You must either wait for water to return or create a new well.

Problems sometime exist when sand or other sediment is pumped out of a shallow well. This is usually a result of having a foot valve or drop pipe that hangs too low in the well. If a well has worked well for a few months and then begins producing sand or other particles, it can be an indication that the well is caving in. Sometimes a foot valve becomes clogged under these conditions. The simple act of shaking the drop pipe can clear a foot valve of debris and allow a pump to return to normal operation.

It's easy to check a shallow well to see if water is in reserve. A weighted line can be dropped into a well to establish the water depth. Assuming that water is present in sufficient quantity, you can rule out a dry well. But, you cannot rule out the fact that the drop pipe or foot valve in the pump might be installed above the water level. This can be checked by pulling the drop pipe out of the well and measuring it. The length of a drop pipe can then be compared to the water depth. If you have water in the well and your drop pipe is submerged, you can rule out the well as your problem.

Drilled wells

Drilled wells rarely run out of water. It is, however, possible that a drilled well could run dry. A weighted line allows you to test for existing water. Pulling the drop pipe and comparing its length to the water depth can prove if the problem is due to the well or the pumping system.

In all of my years as a plumber, I've never known a drilled well to run out of water without some type of outside interference. By outside interference, I mean some form of man-made trouble, such as blasting with explosives somewhere in the general area. Let me give you an example of this from my recent past.

A friend of mine has enjoyed a drilled well for decades. In all of these years, the well had never given its owner any problem until this past summer. In the early summer, road work was being done within a mile or so of my friend's house. Part of the road work involved the blasting of bedrock. Shortly after this blasting took place, my friends well quit producing water. Why? My guess, and it's only a guess, is that the blasting caused a change in the underground water path. It might be that the blasting shifted the rock formations and diverted the water that was serving the well of my friend. I've seen similar situations occur at other times. It's impossible for me to say with certainty that blasting ruined the well, but it's my opinion that it did.

Troubleshooting jet pumps

You are already aware that there are differences between jet pumps and submersible pumps. Knowing this, it only makes sense that there can be differences in the types of problems encountered with the different types of pumps. Let's start our troubleshooting session with jet pumps.

Does not run

A pump that does not run can be suffering from one of many failures (Figs. 16-1 and 16-2). The first thing to check is the fuse or circuit breaker. If the fuse is blown, replace it. When the circuit breaker has tripped, reset it. This is something you could ask your customer to check.

Motor does not start

Cause of trouble	Checking procedure	Correction action
A. No power or incorrect voltage	Using voltmeter check the line terminals. Voltage must be ±10% of rated voltage.	Contact power company if voltage is incorrect.
B. Fuses blown or circuit breakers tripped	Check fuses for recommended size and check for loose, dirty or corroded connections in fuse receptacle. Check for tripped circuit breaker.	Replace with proper fuse or reset circuit breaker.
C. Defective pressure switch.	Check voltage at contact points. Improper contact of switch points can cause voltage less than line voltage.	Replace pressure switch or clean points.
D. Control box malfunction.	For detailed procedure, see pages 29, 30 & 31.	Repair or replace.
E. Defective wiring.	Check for loose or corroded connections. Check motor lead terminals with voltmeter for power.	Correct faulty wiring or connections.
F. Bound pump.	Locked rotor conditions can result from misalignment between pump and motor or a sand bound pump. Amp readings 3 to 6 times higher than normal will be indicated.	If pump will not start with several trials it must be pulled and the cause corrected. New installations should always be run without turning off until water clears.
G. Defective cable or motor.	For detailed procedure, see pages 26, 27 & 28.	Repair or replace.

Motor starts too often

A. Pressure switch.	Check setting on pressure switch and examine for defects.	Reset limit or replace switch.
B. Check valve, stuck open.	Damaged or defective check valve will not hold pressure.	Replace if defective.
C. Waterlogged tank, (air supply)	Check air charging system for proper operation.	Clean or replace.
D. Leak in system.	Check system for leaks.	Replace damaged pipes or repair leaks.

16-1 *Motor troubleshooting chart from a manufacturer's manual.*
A.Y. McDonald MFG. Co., Dubuque, Iowa

When the fuse or circuit breaker is not at fault, check for broken or loose wiring connections. Bad connections account for a lot of pump failures. It is possible the pump won't run due to a motor overload protection device. If the protection contacts are open, the pump won't work. This is usually a temporary condition that corrects itself.

The pump might not run if it is attempting to operate at the wrong voltage. Test the voltage with a volt-ammeter. The power must

Motor runs continuously

Cause of trouble	Checking procedure	Correction action
A. Pressure switch.	Switch contacts may be "welded" in closed position. Pressure switch may be set too high.	Clean contacts replace switch, or readjust setting.
B. Low level well.	Pump may exceed well capacity. Shut off pump, wait for well to recover. Check static and draw-down level from well head.	Throttle pump output or reset pump to lower level. Do not lower if sand may clog pump.
C. Leak in system.	Check system for leaks.	Replace damaged pipes or repair leaks.
D. Worn pump. motor shaft.	Symptoms of worn pump are similar to those of drop pipe leak or low water level in well. Reduce pressure switch setting. If pump shuts off, worn parts may be at fault. Sand is usually present in tank.	Pull pump and replace worn impellers, casing or other close fitting parts.
E. Loose or broken	No or little water will be delivered if coupling between motor and pump shaft is loose or if a jammed pump has caused the motor shaft to shear off.	Check for damaged shafts if coupling is loose and replace worn or defective units.
F. Pump screen blocked.	Restricted flow may indicate a clogged intake screen on pump. Pump may be installed in mud or sand.	Clean screen and reset at less depth. It may be necessary to clean well.
G. Check valve stuck closed.	No water will be delivered if check valve is in closed position.	Replace if defective.
H. Control box malfunction.	See pages 29, 30 & 31 for single phase.	Repair or replace.

Motor runs but overload protector trips

A. Incorrect voltage.	Using voltmeter, check the line terminals. Voltage must be within ± 10% of rated voltage.	Contact power company if voltage is incorrect.
B. Overheated protectors.	Direct sunlight or other heat source can make control box hot causing protectors to trip. The box must not be hot to touch.	Shade box, provide ventilation or move box away from heat source.
C. Defective control box.	For detailed procedures, see pages 29, 30 & 31.	Repair or replace.
D. Defective motor or cable.	For detailed procedures, see pages 26, 27 & 28.	Repair or replace.
E. Worn pump or motor.	Check running current. See pages 11, 14 & 15.	Replace pump and/or motor.

16-2 *Motor troubleshooting chart from a manufacturer's manual.*
A.Y. McDonald MFG. Co., Dubuque, Iowa

be on when this test is conducted. With the leads attached to the meter and the meter set in the proper voltage range, touch the black lead to the white wire, and the red lead to the black wire in the disconnect box near the pump. Test both the incoming and outgoing wiring (Figs. 16-3 and 16-4).

Your next step in the testing process should be at the pressure switch. The black lead should be placed on the black wire, and the red lead should be put on the white wire for this test. There should be a plate on the pump that identifies the proper working voltage. Your test should reveal voltage that is within 10 percent of the recommended rating.

An additional problem that you might encounter is a pump that is mechanically bound. You can check this by removing the end cap and turning the motor shaft by hand. It should rotate freely.

A bad pressure switch can cause a pump to not run. With the cover removed from the pressure switch you can see two springs, one tall and one short. These springs are depressed and held in place by individual nuts. The short spring is preset at the factory and should not be adjusted. (Fig. 16-5).

The long spring can be adjusted to change the cut-in and cut-out pressure for the pump. If you want to set a higher cut-in pressure,

---

Preliminary tests—all sizes—single and three phase

What is to be done	What it means
Measure resistance from any cable to ground (insulation resistance)	1. If the ohm value is normal, the motor windings are not grounded and the cable insulation is not damaged.
	2. If the ohm value is below normal, either the windings are grounded or the cable insulation is damaged. Check the cable at the well seal as the insulation is sometimes damaged by being pinched.
Measure winding resistance (resistance between leads)	1. If all ohm values are normal, the motor windings are neither shorted nor open, and the cable colors are correct.
	2. If any one ohm value is less than normal, the motor is shorted.
	3. If any one ohm value is greater than normal, the winding or the cable is open, or there is a poor cable joint or connection.
	4. If some ohm values are greater than normal and some less on single phase motors, the leads are mixed.

16-3 *Motor troubleshooting chart from a manufacturer's manual.*
A.Y. McDonald MFG. Co., Dubuque, Iowa

Normal ohm and megohm values between all leads and ground

Insulation resistance varies very little with rating. Motors of all hp, voltage, and phase rating have similar values of insulation resistance.

Condition of motor and leads	Ohm value	Megohm value
A new motor (without drop cable).	20,000,000 (or more)	20.0 (or more)
A used motor which can be reinstalled in the well.	10,000,000 (or more)	10.0 (or more)
Motor in well. Ohm readings are for drop cable plus motor.		
A new motor in the well.	2,000,000 (or more)	2.0 (or more)
A motor in the well in reasonably good condition.	500,000–2,000,000	0.5–2.0
A motor which may have been damaged by lightning or with damaged leads. Do not pull the pump for this reason.	20,000–500,000	0.02–0.5
A motor which definitely has been damaged or with damaged cable. The pump should be pulled and repairs made to the cable or motor replaced. The motor will not fail for this reason alone, but it will probably not operate for long.	10,000–20,000	0.01–0.02
A motor which has failed or with completely destroyed cable insulation. The pump must be pulled and the cable repaired or the motor replaced.	less than 10,000	0–0.01

16-4 *Resistance readings.* A.Y. McDonald MFG. Co., Dubuque, Iowa

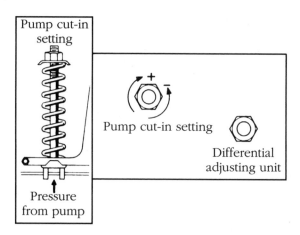

16-5 *Fine-tuning instructions for pressure switches.*
A.Y. McDonald MFG. Co., Dubuque, Iowa

turn the nut tighter to depress the spring further. To reduce the cut-in pressure you should loosen the nut to allow more height on the spring. If the pressure switch fails to respond to the adjustments, it should be replaced.

It is also possible that the tubing or fittings on the pressure switch are plugged. Take the tubing and fittings apart and inspect them. Remove any obstructions and reinstall them.

The last possibility for the pump failure is a bad motor (Figs. 16-6, 16-7, and 16-8). An ohmmeter is used to check the motor. The power to the pump must be turned off.

Meter connections for motor testing

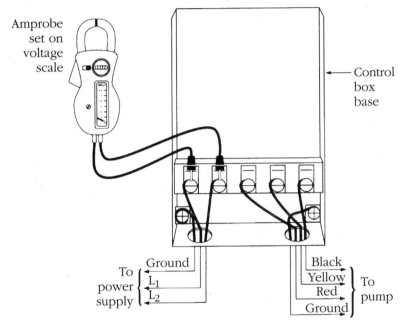

To check voltage

1. Turn power OFF

2. Remove QD cover to break all motor connections.

Caution: L1 and L2 are still connected to the power supply.

3. Turn power ON.

4. Use voltmeter as shown.

Caution: Both voltage and current tests require live circuits with power ON.

16-6 *Meter connections for motor testing.* A.Y. McDonald MFG. Co., Dubuque, Iowa

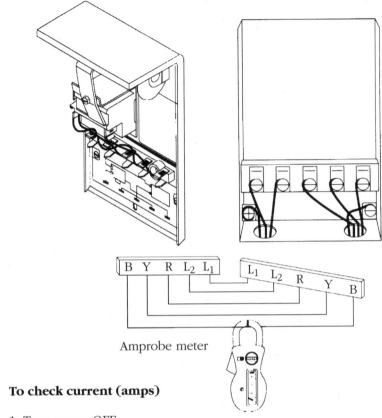

Amprobe meter

To check current (amps)

1. Turn power OFF

2. Connect test cord as shown.

3. Turn power ON.

4. Use hook-on type ammeter as shown.

16-6 *Continued.*

The instructions that follow apply to Goulds pumps with motors rated at 230 volts. When you are conducting the test on different types of pumps, you should refer to the manufacturer's recommendations. Start checking the motor by disconnecting the motor leads. We are going to call these leads L1 and L2.

Set the ohmmeter to R × 100 and adjust the meter to zero. Put one of the meter's leads on a ground screw. The other lead should systematically be touched to all terminals on the terminal board, switch,

Single-phase control boxes
Checking and repairing procedures
(Power on)

Caution: Power must be on for these tests. Do not touch any live parts.

A. General procedures:
1. Establish line power.
2. Check no load voltage (pump not running).
3. Check load voltage (pump running).
4. Check current (amps) in all motor leads.

B. Use of volt/amp meter:
1. Meter such as Amprobe Model RS300 or equivalent may be used.
2. Select scale for voltage or amps depending on tests.
3. When using amp scales, select highest scale to allow for inrush current, then select for midrange reading.

C. Voltage measurements:
   Step 1, no load.
1. Measure voltage at L1 and L2 of pressure switch or line contactor.
2. Voltage Reading: Should be ±10% of motor rating.
   Step 2, load.
1. Measure voltage at load side of pressure switch or line contactor with pump running.
2. Voltage Reading: Should remain the same except for slight dip on starting.

D. Current (amp) measurements:
1. Measure current on all motor leads. Use 5 conductor test cord for Q.D. control boxes.
2. Amp Reading: Current in Red lead should momentarily be high, then drop within one second to values on page 11. This verifies relay or solid state relay operation. Current in Black and Yellow leads should not exceed values on page 11.

E. Voltage symptoms:
1. Excessive voltage drop on starting.
2. Causes: Loose connections, bad contacts or ground faults, or inadequate power supply.

F. Current symptoms:
1. Relay or switch failures will cause Red lead current to remain high and over-load tripping.
2. Open run capacitor(s) will cause amps to be higher than normal in the Black and Yellow motor leads and lower than normal or zero amps in the Red motor lead.
3. Relay chatter is caused by low voltage or ground faults.
4. A bound pump will cause locked rotor amps and overloading tripping.
5. Low amps may be caused by pump running at shutoff, worn pump or stripped splines.
6. Failed start capacitor or open switch/relay are indicated if the red lead current is not momentarily high at starting.

16-7 _Motor troubleshooting chart from a manufacturer's manual._
A.Y. McDonald MFG. Co., Dubuque, Iowa

Single-phase control boxes

Checking and repairing procedures (Power off)

Caution: Turn power off at the power supply panel and discharge capacitors before using ohmmeter.

A. General procedures:
1. Disconnect line power.
2. Inspect for damaged or burned parts, loose connections, etc.
3. Check against diagram in control box for misconnections.
4. Check motor insulation and winding resistance.

B. Use of ohmmeter:
1. Ohmmeter such as Simpson Model 372 or 260. Triplet Model 630 or 666 may be used.
2. Whenever scales are changed, clip ohmmeter lead together and "zero balance" meter.

C. Ground (insulation resistance) test:
1. Ohmmeter Setting: Highest scale R × 10K, or R × 100K
2. Terminal Connections: One ohmmeter lead to "Ground" terminal or Q.D. control box lid and touch other lead to the other terminals on the terminal board.
3. Ohmmeter Reading: Pointer should remain at infinity (∞).

Additional tests

Solid state capacitor run (CRC) control box

A. Run capacitor
1. Meter setting: R × 1,000
2. Connections: Red and Black leads
3. Correct meter reading: Pointer should swing toward zero, then drift back to infinity.

B. Inductance coil
1. Meter setting: R × 1
2. Connections: Orange leads
3. Correct meter reading: Less than 1 ohm.

C. Solid state switch
   Step 1 triac test
1. Meter setting: R × 1,000
2. Connections: R(Start) terminal and Orange lead on start switch.
3. Correct meter reading: Should be near infinity after swing.
   Step 2 coil test
1. Meter setting: R × 1
2. Connections: Y(Common) and L2.
3. Correct meter reading: Zero ohms

16-8 *Motor troubleshooting chart.* A.Y. McDonald MFG. Co., Dubuque, Iowa

capacitor, and protector. If the needle on your ohmmeter doesn't move as these tests are made, the ground check of the motor is okay.

The next check to be conducted is for winding continuity. Set the ohmmeter to R × 1 and adjust it to zero. You need a thick piece of pa-

Ohmmeter tests
Quick disconnect (QD)
solid state control box

A. Start capacitor
1. Meter setting: R × 1,000.
2. Connections: Capacitor terminals.
3. Correct meter reading: Pointer should swing toward Zero, then back to infinity.

B. Solid state switch
Step 1 triac test
1. Meter setting: R × 1,000.
2. Connections: R(Start) terminal and orange lead on start switch.
3. Correct meter reading: Infinity for all models.
Step 2 coil test
1. Meter setting: R × 1.
2. Connections: Y(Common) and L2.
3. Correct meter reading: Zero ohms for all models.

C. Potential (voltage) relay
Step 1 coil test
1. Meter setting: R × 1,000.
2. Connections: #2 & #5.
3. Correct meter readings: For 115 Volt Boxes .7–1.8 (700 to 1,800 ohms). For 230 Volt Boxes 4.5–7.0 (4,500 to 7,000 ohms).
Step 2 contact test
1. Meter setting: R × 1.
2. Connections: #1 & #2.
3. Correct meter reading: Zero for all models.

D. Current relay
Step 1 coil test
1. Meter setting: R × 1.
2. Connections: #1 & #3.
3. Correct meter reading: Less than 1 ohm for all models.
Step 2 contact test
1. Meter setting: R × 1,000.
2. Connections: #2 & #4.
3. Correct meter reading: Infinity for all models.

E. QD (blue) relay
Step 1 triac test
1. Meter setting: R × 1000.
2. Connections: Cap and B terminal.
3. Correct meter reading: Infinity for all models.
Step 2 coil test
1. Meter setting: R × 1.
2. Connections: L1 and B.
3. Correct meter reading: Zero ohms for all models.

16-8 *Continued.*

per for this test. Place the paper between the motor switch points, and the discharge capacitor.

You should read the resistance between L1 and A to see that it is the same as the resistance between A and yellow. The reading between yellow to red should be the same as L1 to the same red terminal (Figs. 16-9 and 16-10).

The next test is for the contact points of the switch. Set the ohmmeter to R × 1 and adjust it to zero. Remove the leads from the switch and attach the meter leads to each side of the switch. You should see a reading of zero. If you flip the governor weight to the run position, the reading on your meter should be infinity (Figs. 16-11 and 16-12).

Now let's check the overload protector. Set your meter to R × 1 and adjust it to zero. With the overload leads disconnected, check the resistance between terminals one and two and then between two and three. If a reading of more than 0.5 occurs, replace the overload protector.

The capacitor can also be tested with an ohmmeter. Set the meter to R × 1000 and adjust it to zero. With the leads disconnected from the capacitor, attach the meter leads to each terminal. When you do this, you should see the meter's needle go to the right and drift slowly to the left. To confirm your reading, switch positions with the meter leads and see if you get the same results. A reading that moves toward zero or a needle that doesn't move at all indicates a bad capacitor.

I realize the instructions I've just given you might seem quite complicated. In a way, they are. Pump work can be very complex. I recommend that you leave major troubleshooting to the person who installed your problem pump. If you are not familiar with controls, electrical meters, and working around electrical wires, you should not attempt many of the procedures I am describing. The depth of knowledge I'm providing might be deeper than you ever expect to use, but it is here for you if you need it.

Runs but gives no water

When a pump runs but gives no water, you have seven possible problems to check out. Let's take a look at each troubleshooting phase in their logical order.

The first consideration should be that of the pump's prime. If the pump or the pump's pipes are not completely primed, water is not going to be delivered. For a shallow-well pump, you should remove the priming plug and fill the pump completely with water. You might want to disconnect the well pipe at the pump and make sure it is holding water. You could spend considerable time pouring water into a priming hole only to find out the pipe is not holding the water.

For deep-well jet pumps, you must check the pressure-control valves. The setting must match the horsepower and jet assembly that's used, so refer to the manufacturer's recommendations.

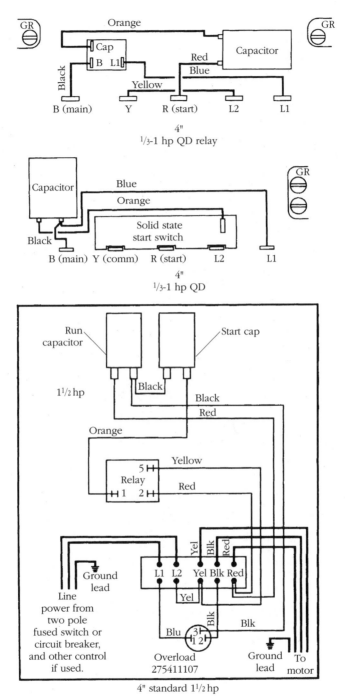

16-9 *Wiring diagrams.* A.Y. McDonald MFG. Co., Dubuque, Iowa

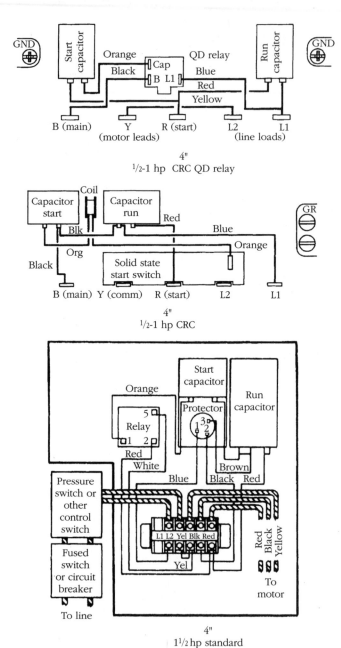

4"
1/2-1 hp CRC QD relay

4"
1/2-1 hp CRC

4"
1 1/2 hp standard

16-9 *Continued.*

Turning the adjustment screw to the left reduces pressure and turning it to the right increase pressure. When the pressure-control valve is set too high, the air-volume control cannot work. If the pressure setting is too low, the pump might shut itself off.

If the foot valve or the end of the suction pipe has become obstructed or is suspended above the water level, the pump cannot produce water. Sometimes shaking the suction pipe can clear the foot valve and get the pump back into normal operation. If you are working with a two-pipe system, you have to pull the pipes and do a visual inspection. However, if the pump you are working on is a one-pipe pump, you can use a vacuum gauge to determine if the suction pipe is blocked.

If you install a vacuum gauge in the shallow-well adapter on the pump, you can take a suction reading. When the pump is running, the gauge does not register any vacuum if the end of the pipe is not below the water level or if there is a leak in the suction pipe.

An extremely high vacuum reading, such as 22 inches or more, indicates that the end of the pipe or the foot valve is blocked or buried in mud. It can also indicate the suction lift exceeds the capabilities of the pump.

When a pump runs without delivering water, the problem commonly is a leak on the suction side of the pump. You can pressurize the system and inspect it for these leaks.

The air-volume control can be at fault on a pump that runs dry. If you disconnect the tubing and plug the hole in the pump, you can tell if the air-volume control has a punctured diaphragm. If plugging the pump corrects the problem, you must replace the air-volume control.

Sometimes the jet assembly becomes plugged. When this happens with a shallow-well pump, you can insert a wire through the ½-inch plug in the shallow-well adapter to clear the obstruction. With a deep-well jet pump, you must pull the piping out of the well and clean the jet assembly.

An incorrect nozzle or diffuser combination can result in a pump that runs but produces no water. Check the ratings in the manufacturer's literature to be sure the existing equipment is the proper equipment.

The foot valve or an in-line check valve could be stuck in the closed position. This situation requires a physical inspection and, if necessary, replacement of the faulty part.

Cycles too often

When a pump cycles on and off too often it can wear itself out prematurely. This type of problem can be caused by several types of

Integral horsepower control box parts

Motor rating hp dia.	Control box (1) model no.	Part no. (2)	Capacitors MFD	Volts	Qty.
5–6"	282 2009 202	275 468 117 S	130–154	330	2
		275 479 103 (5)	15	370	2
	282 2009 203	275 468 117 S	130–154	330	2
		155 327 101 R	30	370	1
5–6" DLX	282 2009 303	275 468 117 S	130–154	330	2
		155 327 101 R	30	370	1
7½–6"	282 2019 210	275 468 119 S	270–324	330	1
		275 468 117 S	130–154	330	1
		155 327 109 R	45	370	1
	282 2019 202	275 468 117S	130–154	330	3
		275 479 103 R (5)	15	370	3
	282 2019 203	275 468 117 S	130–154	330	3
		155 327 101 R	30	370	1
		155 328 101 R	15	370	1
7½–6" DLX	282 2019 310	275 468 119 S	270–324	330	1
		275 468 117 S	130–154	330	1
		155 327 109 R	45	370	1
	282 2019 303	275 468 117 S	130–154	330	3
		155 327 101 R	30	370	1
		155 328 101 R	15	370	1
10–6"	282 2029 210	275 468 119 S	270–324	330	2
		155 327 102 R	35	370	2
	282 2029 202	275 468 117 S	130–154	330	4
		275 479 103 R (5)	15	370	5
	282 2029 203	275 468 117 S	130–154	330	4
		155 327 101 R	30	370	2
		155 328 101 R	15	370	1
	282 2029 207	275 468 119 S	270–324	330	2
		155 327 101 R	30	370	2
		155 328 101 R	15	370	1
10–6" DLX	282 2029 310	275 468 119 S	270–324	330	2
		155 327 102 R	35	370	2
	282 2029 303	275 468 117 S	130–154	330	4
		155 327 101 R	30	370	2
		155 328 101 R	15	370	1
	282 2029 307	275 468 119 S	270–324	330	2
		155 327 101 R	30	370	2
		155 328 101 R	15	370	1
15–6" DLX	282 2039 310	275 468 119 S	270–324	330	2
		155 327 109 R	45	370	3
	282 2039 303	275 468 119 S	270–324	330	2
		155 327 101 R	30	370	4
		155 328 101 R	15	370	1

FOOTNOTES:
(1) Lightning arrestor 150 814 902 suitable for all control boxes
(2) S = Start M = Main L = Line R = Run DXL = Deluxe control box with line contactor.
(3) Capacitor and overload assembly.
(4) 2 required
(5) These parts may be replaced as follows:

Old	New
275 479 102	155 328 102
275 479 103	155 328 101
275 479 105	155 328 103
275 481 102	155 327 102

16-10 *Data chart for single-phase motors.* A.Y. McDonald MFG. Co., Dubuque, Iowa

Overload (2) part no.	Relay (7) part no.	Contactor (2) part no.
155 249 102	155 031 6012	
155 249 102	155 031 601	
155 249 102	155 031 601	155 325 102 L
275 411 102 S 275 406 121 M	155 031 601	
155 249 101	155 031 601	
155 249 101	155 031 601	
275 411 102 S 275 406 121 M	155 031 601	155 326 101 L
155 249 101	155 031 601	155 326 101 L
275 406 102 S 155 409 101 M	155 031 601	
155 249 103	155 031 601 (4)	
155 249 103	155 031 601 (4)	
155 409 101	155 031 102 (6)	155 325 102 S
275 406 102 S 155 409 101 M	155 031 601	155 326 102 L
155 249 103	155 031 601 (4)	155 326 102 L
155 409 101	155 031 102 (6)	155 326 102 L 155 325 102 S
275 406 102 S 155 409 102 M	155 031 601	155 429 101 L
155 409 102	155 031 102 (6)	155 429 101 L 155 325 102 S

(6) May be replaced with 155 031 061 heavy duty relay.
(7) For 208 V systems or where line voltage is between 200 V and 210 volts special low voltage relays are required. Use relay part 155 031 103 in place of part 155 031 102 or 155 031 602 in place of 155 031 601. Also use the next larger cable size than 230 V tables for best operation. Boost transformers per page 12 are an alternative to special relays and cable.

Integral horsepower control box parts

Motor rating hp dia.	Control box (1) model no.	Part no.	MFD	Volts	Qty.
			Capacitors		
1½–4"	282 3008 110	275 464 113 S	105–126	220	1
		155 328 102 R	10	370	1
	282 3007 202 or	275 461 107 S	105–126	220	1
	282 3007 102	275 479 102 R (5)	10	370	1
	282 3007 203 or	275 461 107 S	105–126	220	1
	282 3007 103	155 328 102 R	10	370	1
2–4"	282 3018 110	275 464 113 S	105–126	220	1
		155 328 103 R	20	370	1
	282 3018 202	275 464 113 S	105–126	220	1
		275 479 105 R (5)	20	370	1
	282 3018 203 or	275 464 113 S	105–126	220	1
	282 3018 103	155 328 103 R	20	370	1
2–4" DLX	282 3018 310	275 464 113 S	105–126	220	1
		155 328 103 R	20	370	1
	282 3019 103	275 464 113 S	105–126	220	1
		155 328 103 R	20	370	1
3–4"	282 3028 110	275 463 111 S	208–250	220	1
		155 327 102 R	35	370	1
	282 3028 202	275 463 111 S	208–250	220	1
		275 481 102 R (5)	35	370	1
	282 3028 203 or	275 463 111 S	208–250	220	1
	282 3028 103	155 327 102 R	35	370	1
3–4" DLX	282 3028 310	275 463 111 S	208–250	220	1
		155 327 102 R	35	370	1
	282 3029 103	275 463 111 S	208–250	220	1
		155 327 102 R	35	370	1
5–4" & 6"	282 1138 110	275 468 118 S	216–259	330	1
		155 327 101 R	30	370	2
5–4"	282 1139 202	275 468 118 S	216–259	330	1
		275 479 103 R (5)	15	370	4
	282 1139 203 or	275 468 118 S	216–259	330	1
	282 1139 003	155 327 101 R	30	370	2
5–4" & 6" DLX	282 1138 310 or	275 468 118 S	216–259	330	1
	282 1139 310	155 327 101 R	30	370	2
5–4" DLX	282 1139 303 or	275 468 118 S	216–259	330	1
	282 1139 103	155 327 101 R	30	370	2

FOOTNOTES:
(1) Lightning arrestor 150 814 902 suitable for all control boxes
(2) S = Start M = Main L = Line R = Run DLX = Deluxe control box with line contactor.
(3) Capacitor and overload assembly.
(4) 2 required
(5) These parts may be replaced as follows:

Old	New
275 479 102	155 328 102
275 479 103	155 328 101
275 479 105	155 328 103
275 481 102	155 327 102

16-11 *Data chart for single-phase motors.* A.Y. McDonald MFG. Co., Dubuque, Iowa

Overload (2) part no.	Relay (7) part no.	Contactor (2) part no.
275 411 107	155 031 102	
151 496 922 151 033 946 (3)	155 031 102	
151 496 922 151 033 946 (3)	155 031 102	
275 411 107 S 275 411 113 M	155 031 102	
275 411 107 S 275 411 112 M	155 031 102	
275 411 107 S 275 411 113 M	155 031 102	
275 411 107 S 275 411 113 M	155 031 102	155 325 102 L
275 411 107 S 275 411 102 M	155 031 102	155 325 102 L
275 411 108 S 275 411 115 M	155 031 102	
275 411 108 S 275 406 120 M	155 031 102	
275 411 108 S 275 406 120 M	155 031 102	
275 411 108 S 275 411 115 M	155 031 102	155 325 102 L
275 411 108 S 275 406 120 M	155 031 102	155 325 102 L
275 411 102 S 275 406 102 M	155 031 102	
275 411 102 S 275 406 102 M	155 031 102 (6)	
275 411 102 S 275 406 102 M	155 031 102 (6)	
275 411 102 S 275 406 102 M	155 031 102	155 325 102 L
275 411 102 S 275 406 102 M	155 031 102 (6)	155 325 102 L

(6) May be replaced with 155 031 601 heavy duty relay.
(7) For 208 V systems or where line voltage is between 200 V and 210 volts special low voltage relays are required. Use relay part 155 031 103 in place of part 155 031 102 or 155 031 602 in place of 155 031 601. Also use the next larger cable size than 230 V tables for best operation. Boost transformers per page 12 are an alternative to special relays and cable.

16-11 *Continued.*

QD control box parts

Hp	Volts	Control box model no.	(1) Solid state SW or QD (blue) relay	Start capacitor	MFD
⅓	115	2801024910	152138905(5)	275464125	159–191
		2801024915	223415905(5)	275464125	159–191
⅓	230	2801034910	152138901(5)	275464126	43–53
		2801034915	223415901(5)	275464126	43–53
½	115	2801044910	152138906(5)	275464201	250–300
		2801044915	223415906(5)	275464201	250–300
½	230	2801054910	152138902(5)	275464105	59–71
		2801054915	223415902(5)	275464105	59–71
½	230	2824055010	152138912	275470115	43–52
		2824055015	223415912(6)	275464105	59–71
¾	230	2801074910	152138903(5)	275464118	86–103
		2801074915	223415903(5)	275464118	86–103
¾	230	2824075010	152138913	275470114	108–130
		2824075015	223415913(6)	275470114	86–103
1	230	2801084910	152138904(5)	275464113	105–126
		2801084915	223415904(5)	275464113	105–126
1	230	2824085010	152138914	275470114	108–130
		2824085015	223415914(6)	275470114	108–130

Footnotes:
(1) Prefixes 152 are solid state switches. Prefixes 223 are QD (Blue) Relays.
(2) Control boxes supplied with solid state relays are designed to operate on normal 230 V systems. For 208 V systems or where line voltage is between 200 V use the next larger cable size, or use boost transformer to raise the voltage to 230 V.
(3) Voltage relay kits 115 V, 305 102 901 and 230 V. 305 102 902 will replace either current voltage or QD Relays, and solid state switches.
(4) QD control boxes produced H85 or later do not contain an overload in the capacitor. On winding thermal overloads were added to three-wire motors rated ⅓–1 hp in A85. If a control box dated H85 or later is applied with a motor dated M84 or earlier, overload protection can be provided by adding an overload kit to the control box.
(5) May be replaced with QD relay kits 305 101 901 thru 906. Use same kit suffix as switch or relay suffix.
(6) Replace with CRC QD Relaying Kits, 223 415 912 with 305 105 901, 223 415 913 with 305 105 902 and 223 415 914 with 305 105 903.

16-12 *Data chart for single-phase motors.* A.Y. McDonald MFG. Co., Dubuque, Iowa

problems. For example, any leaks in the piping or pressure tank would cause frequent cycling of the pump.

The pressure switch might be responsible for a pump that cuts on and off too often. If the cut-in setting on the pressure gauge is set too high, the pump works harder than it should.

Volts	Run capacitor	MFD	Volts	Coil
115				
115				
220				
220				
125				
125				
220				
220				
220	155328101	15	370	155662901
220	155327107	15	370	
220				
220				
220	155327108	23	370	155662901
220	155327108	23	370	
220				
220				
220	155327108	23	370	155662901
220	155327108	23	370	

Overload kits

Hp	Volts	Kit number (1)	Kit number (2)
⅓	115	305100901	305091 901
⅓	230	305100902	305091 902
½	115	305100903	305091 903
½	230	305100904	305091 904
¾	230	305100905	305091 905
1	230	305100906	305091 906

(1) For Control Boxes with model numbers that end with 915.

(2) For Control Boxes with model numbers that end with 910.

16-12 *Continued.*

If the pressure tank becomes waterlogged (filled with too much water and not enough air) the pump is going to cycle frequently. If the tank is waterlogged, it must be recharged with air. This would also lead you to suspect a defective air-volume control.

An insufficient vacuum could cause the pump to run too often. If the vacuum does not hold at 3 inches for 15 seconds, it might be the problem.

The last thing to consider is the suction lift. It's possible that the pump is getting too much water and creating a flooded suction. This can be remedied by installing and partially closing a valve in the suction pipe.

Won't develop pressure

Sometimes a pump produces water but it does not build the desired pressure in the holding tank. Leaks in the piping or pressure tank can cause this condition to occur. The same problem might result if the jet or the screen on the foot valve is partially obstructed.

A defective air-volume control might prevent the pump from building suitable pressure. You can test for this by removing the air-volume control and plugging the hole where it was removed. If this solves the problem, you know the air-volume control is bad.

A worn impeller hub or guide vane bore could result in a pump that does not build enough pressure. The proper clearance should be 0.012 inch on a side or 0.025 inch diametrically.

With a shallow-well system, the problem could be caused by the suction lift being too high. You can test for this with a vacuum gauge. The vacuum should not exceed 22 inches at sea level. Deep-well jet pumps require you to check the rating tables to establish their maximum jet depth. Also with deep-well jet pumps, you should check the pressure-control valve to see that it is set properly.

Switch fails

If the pressure switch fails to cut out when the pump has developed sufficient pressure, you should check the settings on the pressure switch. Adjust the nut on the short spring and see if the switch responds. If it doesn't, replace the switch.

Another cause for this type of problem could be debris in the tubing or the fittings that are between the switch and pump. Disconnect the tubing and fittings and inspect them for obstructions.

We have now covered the troubleshooting steps for jet pumps, but before we move on to submersible pumps, I'd like you to look over the illustrations I've given you on engineering data, multi-stage jet pumps, and typical installation procedures.

Troubleshooting submersible pumps

There are major differences between the procedures for troubleshooting submersible pumps and troubleshooting jet pumps. One of the most obvious differences is the fact that jet pumps are installed outside of wells and submersible pumps are installed below the water level of wells.

Sometimes a submersible pump must be pulled out of a well, and this can be quite a chore. Even with today's lightweight well pipe, the strength and endurance needed to pull a submersible pump up from a deep well is considerable. Plumbers that regularly work with submersible pumps often use a pump-puller to make the job easier.

When pulling a submersible pump, you must allow for the long length of the well pipe. This means you must plan the direction from which to pull and also where the pipe and pump are going to lay once removed from the well. It is not unusual to have to deal with between 100 and 200 feet of well pipe, and some wells are even deeper.

When pulling a pump, or lowering one back into a well, it's important to make sure that the electrical wiring does not rub against the well casing. If the insulation on the wiring is cut, the pump is not going to work properly. Let's look now at some specific troubleshooting situations.

Won't start
Pumps that won't start could be the victim of a blown fuse or tripped circuit breaker. If these conditions check out okay, turn your attention to the voltage.

In the following scenarios we are going to be dealing with Goulds pumps and Q-D-type control boxes.

To check the voltage, remove the cover of the control box to break all motor connections. **Caution:** Wires L1 and L2 are still connected to electrical power. These are the wires running to the control box from the power source.

Press the red lead from your voltmeter to the white wire, and the black lead to the black wire. While you are conducting the voltage test, you should turn on any major electrical appliance (such as a clothes dryer) that might be running at the same time as the pump.

Once you have a voltage reading, compare it to the manufacturer's recommended ratings. For example, with a Goulds pump that is rated for 115 volts, the measured volts should range from 105 to 125. A pump with a rating of 208 volts should measure a range from 188 to 228 volts. A pump rated at 230 volts should measure between 210 and 250 volts.

If the voltage checks out okay, check the points on the pressure switch. If the switch is defective, replace it.

The third likely cause of this condition is a loose electrical connection in either the control box, the cable, or the motor. Troubleshooting for this condition requires extensive work with your meters.

To begin the electrical troubleshooting, we are going to look for electrical shorts by measuring the insulation resistance. An ohmmeter is used for this test. The power to the wires you are testing must be turned off.

Set the ohmmeter scale to R × 100K and adjust it to zero. You are going to be testing the wires coming out of the well, from the pump, at the well head. Put one ohmmeter lead to any one of the pump wires and place the other ohmmeter lead on the well casing or a metal pipe. As you test the wires for resistance, you need to know what the various readings mean, so let's examine this issue.

You are going to be dealing with normal ohm values and megohm values. Insulation resistance does not vary with ratings. Regardless of the motor, horsepower, voltage, or phase rating, the insulation resistance remains the same.

A new motor that has not been installed should have an ohm value of 20,000,000 or more and a megohm value of 20. A motor that has been used but is capable of being reinstalled should produce an ohm reading of 10,000,000 or more and a megohm reading of 10.

Once a motor is installed in the well, which is the case in most troubleshooting, the readings are going to be different. A new motor installed with its drop cable should give an ohm reading of 2,000,000 or more and a megohm value of 2.

An installed motor that is in good condition is going to present an ohm reading of between 500,000 and 2,000,000. Its megohm value is going to be between 0.5 and 2.

A motor that gives a reading in ohms of between 20,000 and 500,000 and a megohm reading of between 0.02 and 0.5 might have damaged leads or might have been hit by lightning. However, don't pull the pump yet.

You should pull the pump when the ohm reading ranges from 10,000 to 20,000 and the megohm value drops to between 0.01 and 0.02. These readings indicate that a motor or cables are damaged. While a motor in this condition might run, it probably won't run for long.

When a motor has failed completely or the insulation on the cables has been destroyed, the ohm reading is less than 10,000 and the megohm value is between 0 and 0.01.

With this phase of the electrical troubleshooting done, we are ready to check the winding resistance. You need to refer to charts as reference for correct resistance values, and you must make adjustments if you are reading the resistance through the drop cables. I'll explain more about this in a moment.

If the ohm value is normal during your test, the motor windings are not grounded and the cable insulation is intact. When the ohm readings are below normal, you have discovered that either the insulation on the cables is damaged or the motor windings are grounded.

To measure winding resistance with the pump still installed in the well, you have to allow for the size and length of the drop cable. Assuming you are working with copper wire, you can use the following figures to obtain the resistance of cable for each 100 feet in length and ohms per pair of leads:

Cable Size	Resistance
14	0.5150
12	0.3238
10	0.2036
8	0.1281
6	0.08056
4	0.0506
2	0.0318

If aluminum wire is being tested, the readings are going to be higher. Divide the above ohm readings by 0.61 to determine the actual resistance of aluminum wiring.

If you pull the pump and check the resistance for the motor only (not with the drop cables being tested), you use different ratings. You should refer to a chart supplied by the motor manufacturer for the proper ratings.

When all the ohm readings are normal, the motor windings are fine. If any of the ohm values are below normal, the motor is shorted. An ohm value that is higher than normal indicates that the winding or cable is open or that there is a poor cable joint or connection. Should you encounter some ohm values that are higher than normal while others are lower than normal, you have found a situation where the motor leads are mixed up and need to be attached in their proper order.

If you want to check an electrical cable or a cable splice, you need to disconnect the cable. You also need a container of water. A bathtub can work.

Start by submerging the entire cable, except for the two ends. Set your ohmmeter to R × 100K and adjust it to zero. Put one of the meter leads on a cable wire and the other to a ground. Test each wire in the cable with this same procedure.

If, at anytime, the meter's needle goes to zero, remove the splice connection from the water and watch the needle. A needle that falls back to give no reading indicates that the leak is in the splice.

Once the splice is ruled out, you have to test sections of the cable in a similar manner. In other words, once you have activity on the meter, you should slowly remove sections of the cable until the meter settles back into a no-reading position. When this happens, you have found the section that is defective. At this point, the leak can be covered with waterproof electrical tape and reinstalled, or you can replace the cable.

Doesn't run

A pump that doesn't run can require extensive troubleshooting. Start with the obvious and make sure that the fuse is not blown or that the circuit breaker is not tripped. Also check to see that the fuse is of the proper size.

Incorrect voltage can cause a pump to fail. You can check the voltage as described in the electrical troubleshooting section above.

Loose connections, damaged cable insulation, and bad splices, as discussed above, can prevent a pump from running.

The control box can have a lot of influence on whether or not a pump runs. If the wrong control box has been installed or if the box is located in an area where temperatures rise to over 122 degrees F, the pump might not run.

When a pump does not run, you should check the control box carefully. We are going to be working with a quick-disconnect type box. Start by checking the capacitor with an ohmmeter. First, discharge the capacitor before testing. You can do this by putting the metal end of a screwdriver between the capacitor's clips. Set the meter to R × 1000 and connect the leads to the black and orange wires out of the capacitor case. You should see the needle start towards zero and then swing back to infinity. Should you have to recheck the capacitor, reverse the ohmmeter leads.

The next check involves the relay coil. If the box has a potential relay (3 terminals), set your meter on R × 1000 and connect the leads to red and yellow wires. The reading should be between 700 and 1800 ohms for 115-volt boxes. A 230-volt box should read between 4500 and 7000 ohms.

If the box has a current relay coil (4 terminals), set the meter on R × 1 and connect the leads to black wires at terminals one and three. The reading should be less than 1 ohm.

In order to check the contact points you must set your meter on R × 1 and connect to the orange and red wires in a 3-terminal box.

The reading should be zero. For a 4-terminal box, you must set the meter at R × 1000 and connect to the orange and red wires. The reading should be near infinity.

Now you are ready to check the overload protector with your ohmmeter. Set the meter at R × 1, and connect the leads to the black wire and the blue wire. The reading should be at a maximum of 0.5.

If you are checking the overload protector for a control box designed for 1½ horsepower or more, you are going to set your meter at R × 1 and connect the leads to terminal number one and to terminal number three on each overload protector. The maximum reading should not exceed 0.5 ohms.

A defective pressure switch or an obstruction in the tubing and fittings for the pressure switch could cause the pump not to run.

As a final option, the pump might have to be pulled and checked to see if it is bound. There should be a high amperage reading if this is the case.

Doesn't produce water

When a submersible pump runs but doesn't produce water, there are several things that could be wrong. The first thing to determine is if the pump is submerged in water. If you find that the pump is submerged, you must begin your regular troubleshooting.

Loose connections or wires connected incorrectly in the control box could be at fault. The problem could be related to the voltage. A leak in the piping system could easily cause the pump to run without producing adequate water.

A check valve could be stuck in the closed position. If the pump was just installed, the check valve might be installed backwards. Other options include a worn pump or motor, a clogged suction screen or impeller, and a broken pump shaft or coupling. You must pull the pump if any of these options are suspected.

Tank pressure

If you don't have enough tank pressure, check the setting on the pressure switch. If that's okay, check the voltage. Next, check for leaks in the piping system, and as a last resort, check the pump for excessive wear.

Frequent cycling

Frequent cycling is often caused by a waterlogged tank, as was described in the section on jet pumps. Of course, an improper setting on the pressure switch can cause a pump to cut on too often, and leaks in the piping can be responsible for the trouble. You might find the problem is caused by a check valve that has stuck in an open position.

A pressure tank that is sized improperly can cause problems. The tank should allow a minimum of one minute of running time for each cycle.

Water quality

Water quality in private water supplies can vary greatly from one building lot to another. It can even fluctuate within one water source. A well that tests fine one month might test differently six months later. Frequent testing is the only way to assure a good quality of water. Many times some type of treatment is either desirable or needed before water can be used for domestic purposes.

Many people have heard of hard water and soft water. While these names might be familiar, a lot of people don't know the differences between the two types. Some water supplies have hazardous concentrations of contaminants, but most wells are affected more often by mineral contents. While the mineral contents might not necessarily be harmful to a person's health, they can create other problems.

Is it a builder's responsibility to provide customers with water that is free of mineral content? It shouldn't be. Water conditioning equipment can get very expensive. Spending $1500, or more, for such equipment would not raise any eyebrows in the plumbing community. If you want to make sure that you don't become entangled in litigation over water quality, have your lawyer create a disclaimer clause for your contracts.

Four prime characteristics can affect the quality of water. Physical characteristics are the first. By physical characteristics, we are talking about such factors as color, turbidity, taste, odor, and so forth. Chemical differences create a second category. This type of situation can involve such topics as hard and soft water. Third, biological contents can modify both the physical and chemical characteristics of water. Biological contents can often render water unsafe for drinking. The fourth consideration is radiological factors, such as radon. To understand these four groups better, let's look at them individually.

Physical characteristics

The physical characteristics of water provide one way to assess water quality. Taste and odor rank high as concerns in the category of physical characteristics. Water that doesn't taste good or that smells bad is undesirable. It might be perfectly safe to drink, but its physical qualities make it unpleasant.

What causes problems with taste and odors? Foreign matter is at fault. It might come in the form of organic compounds, inorganic salts, or dissolved gases. A sulfur content is one of the best known causes of odor in water.

Water that doesn't look good can be difficult to drink enjoyably. Color is a physical characteristic that can taint someone's opinion of water quality. Having colored water rarely indicates a health concern, but it is a sure way of drawing some customer complaints. Dissolved organic matter from decaying vegetation and certain inorganic matter typically give water its color.

Turbidity
Do you know what turbid water is? It's simply water that is cloudy in appearance. Turbidity is caused by suspended materials in the water. Such materials might be clay, silt, very small organic material, plankton, and inorganic materials. Turbidity doesn't usually pose a health risk, but it makes water distasteful to look at in a drinking glass.

Temperature
Technically, water temperature falls into the category of physical characteristics. This, however, is not a problem that many people complain about. Deep wells produce water at consistent temperatures. Shallow wells are more likely to have water temperatures that fluctuate. In either case, temperature is rarely a concern with wells.

Foam
Foam is something that you do not normally encounter in well water. However, foamability is a physical characteristic of water, and it can be an indicator that serious water problems exist. Water that foams is usually affected by some concentration of detergents. While the foam itself might not be dangerous, the fact that detergents are reaching a water source should raise some alarm. If detergents can invade the water, more dangerous elements might also be present.

Chemical characteristics
Chemical characteristics are often monitored in well water. A multitude of chemical solutions might be present in a well. As long as quantities are within acceptable, safe guidelines, the presence of chemicals does not automatically require action. However, concentrations of some chemicals can prove harmful. The following is a list

of some chemicals that might be present in the next well you have installed:

- Arsenic
- Barium
- Cadmium
- Chromium
- Cyanides
- Fluoride
- Lead
- Selenium
- Silver

Chlorides

Chlorides, in solution, are often present in well water. If an excessive quantity of chlorides is present, it might indicate pollution of the water source.

Copper

Copper can be found in some wells as a natural element. Aside from giving water a poor taste, small amounts of copper are not usually considered harmful.

Fluorides

Why pay a dentist to give your children fluoride treatments when fluorides might be present in your drinking water. Fluorides can be found naturally in some well water. Too much fluoride in drinking water is not good for teeth, so quantities should be measured and assessed by experts.

Iron

Iron is commonly found in well water. If water has a high iron content, it is difficult to avoid getting brown stains on freshly-washed laundry. Also, plumbing fixtures can be stained by water containing too much iron. The taste of water containing iron can be objectionable.

Lead

You don't want lead to show up on a well test. It's possible for dangerous levels of lead to exist in a water source. However, lead found in water generally can be traced to the use of lead plumbing pipes. Modern plumbing codes prohibit lead pipes, but older homes don't share this safeguard.

Manganese

Manganese, like iron, is very common in well water. Staining of laundry and plumbing fixtures is one reason to limit the amount of manganese in domestic water. It is not unusual for manganese to adversely affect the taste of water. Excessive consumption of manganese can cause health problems.

Nitrates

Nitrates show up most often in shallow wells. When infants drink water that contains nitrites, it can cause what is known as blue-baby disease. Shallow wells that are located near livestock are susceptible to nitrate invasion.

Pesticides

As we all know, pesticides and well water shouldn't mix. Shallow wells located in areas where pesticides are used should be regularly checked to confirm the suitability of the water for domestic use. Many reports indicate that wells have been contaminated during ground treatments for termite control.

Sodium

Sodium can show up in well water. For average, healthy people, this is not a problem. However, individuals who are forced to maintain low-sodium diets can be affected by the sodium content in well water.

Sulfates

Sulfates in well water can act as a natural laxative. You can imagine why this condition would not be desirable in most homes.

Zinc

Zinc doesn't normally draw attention to itself as a health risk, and it is not a common substance in well water. But, it can sometimes be present. Taste is normally the only objection to having zinc in well water.

Hard water

Hard water is a common problem among well users. This water doesn't work well with soap and detergents. If you heat a pot of water on a stove and find a coating of a white dust-like substance left in the pan, it is a strong indication that hard water is present. This same basic coating can attack plumbing pipes and storage tanks, creating a number of plumbing problems. Hard water can even cause flush holes in the rims of toilets to clog, making the toilets flush slowly or poorly.

Acidic water

Acidic water is not uncommon in wells. When water has a high acid content, it can eat holes through the copper tubing used in plumbing systems. Plumbing fixtures can be damaged from acidic water. People with sensitive stomachs can suffer from a high acid content. The acidity of water is measured on a pH scale. This scale runs from zero to 14. A reading of seven indicates neutral water. Any reading below seven is moving into an acidic range. Numbers above seven are inching into an alkaline status.

Biological factors

Biological factors can be a big concern when it comes to well water. For water to be potable, it must be free of disease-producing organisms. What are some of the organisms? Bacteria, especially of the coliform group, are one of them. Others include protozoa, virus, and helminths (worms).

Biological problems can be avoided in many ways. One way is to use a water source that doesn't support much plant or animal life, such as a well. Springs, ponds, lakes, and streams are more likely to produce biological problems than covered wells.

It is also necessary to protect a potable water source from contamination. The casing around a well does this. Light should not be able to shine on a water source, and a well cap or cover meets this requirement. Temperature can also play a part in bacterial growth, but the temperature of most wells is not something to be concerned about. If biological activity is a problem, various treatments to the water can solve the problem.

Radiological factors

Radiological factors are not a big threat to most well users, but some risk does exist. Special testing can be done to determine if radioactive materials are a significant health risk in any given well. This test should, in my opinion, be conducted by experienced professionals. The same goes for biological testing.

Solving water-quality problems

A small book could be written on the many ways to solve water-quality problems. Rather than give you a full tour of all aspects of water treatment, I am going to concentrate on the methods most often used. Some form of treatment is available for nearly any problem you encounter.

Bacteria

Bacteria in well water is a serious problem. The most common method of dealing with this problem doesn't require any fancy treatment equipment. A quantity of chlorine bleach is usually all that is needed. Bleach is added to the contaminated well and allowed to settle for awhile. After a prescribed time, the well is drained or run until no trace of the bleach is evident. This normally clears up biological activity. Sophisticated treatment systems do exist for nasty water, but the odds of needing it are remote.

Acid neutralizers

Acid neutralizers are available at a reasonable cost to control high acid contents in domestic water. These units are fairly small, easy to install, not difficult to maintain, and they don't cost a small fortune.

Iron and manganese

Iron and manganese can be controlled with iron-removal filters. Like acid neutralizers, these units are not extremely expensive, and they can be installed in a relatively small area. If both a water softener and iron-removal system are needed, the iron-removal system should treat water before it reaches a water softener. Otherwise, the iron or manganese might foul the mineral bed in the water softener.

Water softeners

Water softeners can treat hard water and bring it back to a satisfactory condition. The use of such a treatment system can prolong the life of plumbing equipment, while providing users with better water quality.

Activated carbon filters

Activated carbon filters provide a solution for water that has a foul taste or odor. Many of these filters are simple, in-line units that are inexpensive and easy to install. When conditions are severe, a more extensive type of activated carbon filter might be required. This type of filter can remove the ill effects of sulfur water (hydrogen sulfide).

Turbidity

Turbidity can be controlled with simple, in-line filters. If the water contains high amounts of particles, the cartridges in these filters must be changed frequently. Left unchanged, they can collect so many particles that water pressure is greatly reduced.

Well, this has been a long, but informative chapter. I don't know if you are tired of reading, but I'm tired of writing. Let's take a short break before we move to the next chapter where we are going to examine common septic tank problems and their solutions.

17

Routine remedies for sorrowful septic systems

This chapter is filled with routine remedies for sorrowful septic systems. If you install new septic systems, sooner or later you are bound to run into some problems after the job is done. Like any other type of new installation, there is a risk that something can go wrong while the installation is still under warranty. The homeowner's problem becomes the builder's problem.

If a builder subs septic work out to an independent contractor, it does not make the independent contractor solely responsible for problems with the septic system. Since the builder is the general contractor, a customer has a right to look for help from the builder. In other words, even if you don't put a septic system in yourself, you are still responsible for it as the general contractor.

As a general contractor who has subcontracted a septic system out, you are fortunate to have someone to call to solve your septic problems. If you install septic systems with an in-house crew, you or your employees must assume full responsibility. Either way, you need to be able to talk intelligently to your customers when problems arise.

Several types of potential failures need to be considered. Many of them are avoidable. If a builder informs customers of the do's and don'ts when a house is sold, many problems can be avoided. Do you know what recommendations to make to your customers in regards to the use of their septic systems? Many contractors don't. Even the ones that know what should be said often never take the time to inform their customers. To me, this is stupid. If I can avoid warranty work by informing my customers, you can bet I am going to spend a few minutes giving them the facts.

We are about to discuss many potential septic problems. But before we do, let me ask you a question. Pretend you are building a house for me. Am I going to have any problems with my septic tank if you install a garbage disposer under my kitchen sink? Do you know what to tell me? If you don't, you should.

Garbage disposers are often an issue that brings conflicting reports from experts. Some plumbing codes prohibit the use of garbage disposers in houses with septic systems. Other plumbing codes permit such installations. The building code information would be the first thing you would need to know in order to answer my installation question. If it's against code to install a disposer, you have a quick answer to give me.

Assuming that the code allows the use of a disposer, the answer you give me should be more complex. For example, I might need a large septic tank to accommodate the use of a disposer. There is some risk that large chunks of food from the disposer could clog up my drain field. If the field becomes clogged, it has to be dug up and repaired or replaced. It is not an inexpensive undertaking.

How you answer my installation question could have some serious effects on your company. For example, if you have me sign a notice that explains the precautions to take and also the potential risk I'm assuming by requiring you to install a garbage disposer, you're pretty well off the hook. If you don't advise me of the potential risks and simply tell me that it's no problem to install a disposer, I might be able to build a lawsuit against you when my septic system fails. Keep this scenario in mind as we discuss the various potential problems with septic systems. The more you can disclose to a customer in writing, the better off you are. It is, of course, always best to have customers sign a copy of the disclosure and keep it in your files.

An overflowing toilet

Some homeowners associate an overflowing toilet with a problem in their septic system. It is possible that the septic system is responsible for the toilet backing up, but this is not likely in a house that is still under warranty. It is more likely to be a stoppage either in the toilet trap or in the drain pipe. Knowing this can help you decide who to send out on the call.

If you get a call from a customer who has a toilet flooding the bathroom, there is a quick, simple test that the homeowner can perform to tell you more about the problem. You know the toilet won't drain, but how about the kitchen sink? Can other toilets in the house drain? If other fixtures drain just fine, the problem is not with the septic tank.

You should give your customers some special instructions prior to having them test other fixtures. First, it is best if they use fixtures that are not in the same bathroom with the plugged-up toilet. Lavatories and bathing units often share the same main drain that a toilet uses. Testing a lavatory that is near a stopped-up toilet can tell you if the toilet is the only fixture affected. It can, in fact, narrow the likelihood of the problem down to the toilet's trap. But, if the stoppage is some way down the drain pipe, it's conceivable that the entire bathroom group is affected. It is also likely that if the septic tank is the problem, water is going to back up in a bathtub.

When an entire plumbing system is unable to drain, water rises to the lowest fixture, which is usually a bathtub or shower. So, if there is no back-up in a bathing unit, there probably isn't a problem with the septic tank. But, back-ups in bathing units can happen even when the major part of a plumbing system is working fine. A stoppage in a main drain could cause the liquids to back up, into a bathing unit.

To determine if there is a total back-up, have homeowners fill their kitchen sinks and then release all of the water at once. Get them to do this several times. A volume of water might be needed to expose a problem. Simply running the faucet for a short while might not show a problem with the kitchen drain. If the kitchen sink drains successfully after several attempts, it's highly unlikely that there is a problem with the septic tank. This would mean that you should call your plumber, not your septic installer.

Whole-house back-ups

Whole-house back-ups (where none of the plumbing fixtures drain) indicate either a problem in the building drain, the sewer, or the septic system. There is no way to know where the problem is until some investigative work is done. This is not a good job to assign to homeowners. Your plumber is the most logical subcontractor to call when this problem exists. It's possible that the problem is associated with the septic tank, but your plumber can pinpoint the location.

For all the plumbing in a house to back up, there must be some obstruction at a point in the drainage or septic system beyond where the last plumbing drain enters the system. Plumbing codes require clean-out plugs along drainage pipes. There should be a clean-out either just inside the foundation wall of a home or just outside the wall.

This clean-out location and the access panel of a septic tank are the two places to begin a search for the problem.

If the access cover of the septic system is not buried too deeply, I would start there. But, if extensive digging would be required to expose the cover, I would start with the clean-out at the foundation, hopefully on the outside of the house. Your plumber should be able to remove the clean-out plug and snake the drain. This normally clears the stoppage, but you might not know what caused the problem. Habitual stoppages point to a problem in the drainage piping or septic tank.

Removing the inspection cover from the inlet area of a septic tank can show you a lot. For example, you might see that the inlet pipe doesn't have a tee fitting on it and has been jammed into a tank baffle. This could obviously account for some stoppages. Cutting the pipe off and installing the diversion fitting can solve this problem.

Sometimes pipes sink in the ground after they are buried. Pipes sometimes become damaged when a trench is backfilled. If a pipe is broken or depressed during backfilling, there can be drainage problems. When a pipe sinks in uncompacted earth, the grade of the pipe is altered, and stoppages become more likely. You might be able to see some of these problems from the access hole over the inlet opening of a septic tank.

Once you remove the inspection cover of a septic tank, look at the inlet pipe. It should be coming into the tank with a slight downward pitch. If the pipe is pointing upward, it indicates improper grading and a probable cause for stoppages. If the inlet pipe either doesn't exist or is partially pulled out of the tank, there's a very good chance that you have found the cause of your back-up. If a pipe is hit with a heavy load of dirt during backfilling, it can be broken off or pulled out of position. This won't happen if the pipe is supported properly before backfilling, but someone might have cheated a little during the installation.

In the case of a new septic system, a total back-up is most likely to be the result of some failure in the piping system between the house and the septic tank. If your problem is occurring during very cold weather, it is possible that the drain pipe has retained water in a low spot and that the water has frozen. I've seen this happen several times in Maine with older homes (not ones that I've built).

Running a plumber's snake from the house to the septic tank can tell you if the problem is in the piping. This is assuming that the snake used is a pretty big one. Little snakes might slip past a blockage that is capable of causing a back-up. An electric drain-cleaner with a full-size head is the best tool to use.

The problem is in the tank

Sometimes, even with new systems, the problem causing a whole-house back-up is in the septic tank. Such occasions are rare, but they do exist. When this is the case, the top of the septic tank must be uncovered. Some tanks, like the one at my house, are only a few inches beneath the surface. Other tanks can be buried several feet below the finished grade. If you built the house recently, you should know where the tank is located and how deeply it is buried.

Once a septic tank is in full operation, it works on a balance basis. The inlet opening of a septic tank is slightly higher than the outlet opening. When water enters a working septic tank, an equal amount of effluent leaves the tank. This maintains the needed balance. But, if the outlet opening is blocked by an obstruction, water can't get out. This causes a back-up.

Strange things sometimes happen on construction sites, so don't rule out any possibilities. It might not seem logical that a relatively new septic tank could be full or clogged, but don't bet on it. I can give you all kinds of things to think about. Suppose your septic installer was using up old scraps of pipe for drops and short pieces, and didn't notice that one of the pieces had a plastic test cap glued on the end? This could certainly render the septic system inoperative once the liquid rose to the level of the capped outlet drain. Could this really happen? I've seen the same type of situation with interior plumbing, so it could happen at a septic tank.

What else could block the outlet of a new septic tank? Maybe a piece of scrap wood found its way into the septic tank during construction and is now blocking the outlet. The point is, almost anything could be happening in the outlet opening, so take a snake and see if it is clear.

If the outlet opening is free of obstructions, and all drainage to the septic tank has been ruled out as a potential problem, you must look farther down the line. Expose the distribution box and check it. Run a snake from the tank to the box. If it comes through without a hitch, the problem is somewhere in the leach field. In many cases, a leach field problem can cause the distribution box to flood. So, if you have liquid rushing out of the distribution box, you should be alerted to a probable field problem.

Problems with a leach field

Problems with a leach field are uncommon among new installations. Unless the field was poorly designed or improperly installed, there is very little reason why it should fail. However, extremely wet ground conditions, due to heavy or constant rains, could force a field to become saturated. If the field saturates with groundwater, it cannot accept the effluent from a septic tank. This, in turn, causes back-ups in houses. When this is the case, the person who created the septic design might be at fault.

I've never built a house that had a failure of the septic system field. But, as a plumbing contractor, I've responded to such calls, even with fairly new houses. The problem is usually a matter of poor workmanship. If you keep a watchful eye on your septic crew during the installation, you should not be awakened in the middle of the night by an irate customer with a failed leach field.

Assuming that you are unfortunate enough to experience a saturated drain field, your options are limited. You might wait a few days in hopes that the field dries out. Beyond that option, you've got some digging to do. Extending the size of the drain field is the only true solution when groundwater saturation is a seasonal problem.

Older fields

Older fields often clog up and fail. This problem is not likely to happen while a field is under warranty. But, you might be interested to learn the types of situations that can cause a field failure.

Clogged with solids

Some drain fields become clogged with solids. Financially, this is a devastating discovery. A clogged field has to be dug up and replaced. Much of the crushed stone might be salvageable, but the pipe, the excavation, and any new stone can cost thousands of dollars. The reasons for a problem of this nature is either poor design, bad workmanship, or abuse.

If the septic tank installed for a system is too small, solids are likely to enter the drain field. An undersized tank could be the result of a poor septic design, or it could come about as a family grows and adds to their home. A tank that is adequate for two people might not be satisfactory for four people. Unfortunately, finding out that a tank is too small often doesn't happen until the damage has already been done.

As a builder, you might take on jobs that involve the building of room additions. If you do, and if a septic system is involved, you should check to see that you are not putting too much burden on the septic system. Local officials might require upgrades to an existing septic system when bedrooms are added to a house.

Why would a small septic tank create problems with a drain field? Septic tanks accept solids and liquids. Ideally, only liquids should leave the septic tank and enter the leach field. Bacterial action occurs in a septic tank to break down solids. If a tank is too small, there is not adequate time for the breakdown of solids to occur. Increased loads on a small tank can force solids into the drain field. After this happens for awhile, the solids plug up the drainage areas in the field. This is when digging and replacement is needed.

Too much pitch

Is there any such thing as having too much pitch on a drain pipe? Yes, there is. A pipe that is graded with too much pitch can cause several problems. In interior plumbing, a pipe with a fast pitch might allow water to race by without removing all the solids. A properly graded pipe floats the solids in the liquid as drainage occurs. If the water is allowed to rush out, leaving the solids behind, a stoppage eventually occurs.

In terms of a septic tank, a pipe with a fast grade can cause solids to be stirred up and sent down the outlet pipe. When a four-inch wall of water dumps into a septic tank at a rapid rate, it can create quite a ripple effect. The force of the water might generate enough stir to float solids that should be sinking. If these solids find their way into a leach field, clogging is likely.

Garbage disposers

We talked a little bit about garbage disposers earlier. A garbage disposer adds more solids to a septic system. Because of the added solids, a larger tank is needed for satisfactory operation and a reduction in the risk of a clogged field. I remind you again, some plumbing codes prohibit the use of garbage disposers with septic systems.

Other causes

Other causes for field failures can be related to collapsed piping. This is not common with today's modern materials, but it is a fact of life with some old drain fields. Heavy vehicular traffic over a field can compress it and cause the field to fail. This is true even of modern fields. Also, saturation of a drain field can cause it to fail. This could be the result of seasonal water tables or prolonged use of a field that is giving up the ghost.

Septic tanks should have the solids pumped out of them on a regular basis. For a normal residential system, pumping once every two years should be adequate. Septic professionals can measure sludge levels and determine if pumping is needed. Failure to pump a system routinely can result in a build-up of solids that could invade and clog a leach field.

My house stinks

Have you ever had a customer call and say that their house stinks? This has never happened to me as a builder, but my plumbing company has dealt with this complaint. Houses sometimes develop unpleasant odors associated with a drainage or septic system. This can occur even in new houses. In fact, it happened in my mother-in-law's new house a couple of years ago.

Sewer gas occurs naturally in drains and septic systems. The odor associated with this gas, which is extremely flammable, is normally controlled by using vent pipes and water-filled fixture traps. If a fixture trap loses its water seal, gas can escape into a home. This is dangerous and unpleasant. A faulty wax ring under a toilet can lead to the same problem. So, too, can blocked vents and leaking pipe joints.

Call your plumber if you have a customer who is complaining of sewer odors inside their house. The first thing an experienced plumber checks is the trap seals. Sometimes, when a fixture is not often used, the water evaporates in the fixture trap. When this happens, sewer gas has a direct path into a home. This is common with floor drains and fixtures that are seldom used.

In my mother-in-law's case, the problem was a dry trap at the downstairs shower. The shower was rarely used, and the trap seal evaporated. I ran the shower for a few minutes to fill the trap and the problem was solved. Now she periodically runs water in the shower to prevent a reoccurrence of the odor.

Floor drains are frequently a cause of interior odors. Since these drains see little use, their seals evaporate. Unless the trap is fitted with a trap primer, water must be poured down the drain every now and then. A primer is a little water line that automatically maintains a trap seal, but it is not common in residential applications.

When the source of the odor seems to be a toilet, the wax seal between the toilet bowl and its flange should be replaced. More complex problems in a plumbing system could exist. For example, I once had a new house where the plumbing crew forgot to remove the test caps from the roof vents. Capped off, the vents couldn't work. This affected the drainage more than the odors. A poorly-vented drain is going to drain very slowly. But, a plugged vent, such as one that is filled with ice or a bird's nest, can start an odor problem.

If leaking joints are suspected in a vent or drainage system, there are special tests that your plumber can perform. Colored smoke can be used to reveal leak locations. Another test uses a peppermint smell to pinpoint odor leaks. Your plumber should be aware of both of these tests.

Outside odors

Outside odors normally have to do with a leach field. But, the problem could be with the plumbing vents on top of a house. Sewer gas escapes from plumbing vents. Under most conditions, it goes unnoticed. But, under the right weather conditions, such as heavy air with no breeze, the odor from a vent might be forced down to where it can be smelled. If a vent is too close to an open window, the gas can come into a house. The plumbing code sets standards for vent placement and height. If these regulations are observed, a problem should not exist. However, a short vent pipe or a vent that is close to a window or other ventilating opening could cause some problems.

Puddles and odors

Puddles and odors are sometimes found in and near leach fields. If you have septic puddles, you're going to have septic odors. Pumping out the septic tank won't help here. Your problem is with the field itself. It is not too unusual for this type of problem to attack fairly new systems that were not installed properly.

One main reason for puddles in a new system is the grade on the distribution pipes in the leach field. These pipes should be installed relatively level. If they have a lot of pitch, effluent runs to the low end and builds up. This causes the puddle and the odor. A sloppy installer who doesn't maintain an even, nearly nonexistent grade while installing distribution pipes can be the root cause of your problem. And, it's an expensive problem to solve.

Since the effluent quickly runs to a low spot, most of the absorption field is not being utilized. It doesn't mean that the field is too small, although this could cause liquid to surface. In the case of overgraded pipes, the problem is due to the fact that most of the leach field is not used, and the volume of liquid dumped in the low spot cannot be absorbed quickly enough. To solve this problem, you must excavate the field and correct the pipe grade. This solution is not a cheap one. If, as a builder, you find the field pipes were installed with too much grade, you should have some opportunity to force the septic installer to pay for corrections.

Saturation in part of a septic field can cause outside odors. In time, the ground might drain naturally and solve the problem. This could be the case if odors occur only after heavy or prolonged rains. You shouldn't have outside odors with septic systems. When they are present, especially for several days, you are probably looking at an expensive problem.

Supervision

One of your best protections as a builder is to supervise the work as the septic systems are installed. Homeowners might not expect you to be an expert in the installation of septic systems, but you can bet that they expect you to provide them with a good job. If this means crawling down into a septic bed and putting a level on the distribution pipes to check for excessive grading, then do it. As the builder and general contractor of a house with a septic system, you cannot avoid responsibility for the system.

You can take an attitude that says you shouldn't have to get personally involved with the installation of a septic system. Your position might be that you hire competent professionals to make your septic installations, and a code office inspects the work, so why should you get directly involved? I don't think that this position can hold up in court. Even if you never go to court, you have your reputation as a builder to consider. If your customers are getting bad septic systems, you're going to get a bad reputation. Some supervision on your part can avoid this problem.

When a problem occurs

Take fast action when a problem occurs with a septic system that you had installed. Don't put off your customers. Most people forgive mistakes as long as corrective action is quickly taken. Since you are not likely to know the source of the problem, a call to your plumber is usually a good first step. If you can't get your plumber quickly, go to the customer's house personally. An inspection by you might not reveal the cause of a septic problem, but it is going to have a favorable impact on the customer. A visit of this type also buys you time to get your plumber or septic installer on the job.

A quick response is necessary when a customer's plumbing is failing. However, avoid making hasty decisions on the cause of the problem. Take your time. Make your plumber investigate the situation thoroughly. Solve the problem right the first time. If you make some token gesture to fix the problem and it doesn't work, you are going to look incompetent. It's better to spend enough time to fix a problem right the first time.

Don't be afraid to ask for help

Don't be afraid to ask for help when you are faced with a septic problem. Unless you installed the system yourself, you should have some professionals available to help you. Your plumber and your septic installer can both be of help, and they both have a stake in your problems. Call them. Ask them to come out and investigate the problem. If they like working for you, they are going to come.

Even if you installed a system yourself, consider calling in some experts when you are up against tough troubleshooting problems. Septic installers should be willing to help you for a price. County officials are another potential source of help. The person who drew your septic design might be able to shed some light on the cause of your problem. Make some phone calls. Get advice, even if you have to pay for it. You owe it to your customers to solve their warranty problems. If you don't offer your help freely, you might find yourself being served with legal papers.

Fortunately, problems with new septic systems are rare. Unless someone made a mistake in the design or installation, a septic system should function without failure for many years. A good system could easily go twenty years or more, without anything more than routine pumping. With some exceptions, a leach field should last indefinitely when installed with modern materials and proper workmanship.

Don't be afraid of septic systems. It's okay to respect them, but there is no reason to fear them. I've worked with septic systems for decades, and I've never had any significant problem with one. If you make it a rule not to build houses where septic systems are needed, you are throwing away a lot of good work potential.

We are about to enter our last chapter. It is an important one. If you can't afford cost overruns, you are going to appreciate Chapter 18. It's all about ways to reduce financial losses. Let's turn to it now and get some new ammunition that can help you avoid losing money on future jobs.

18

Making more money by avoiding cost overruns

You can make more money if you avoid cost overruns. I'm sure you know this, but you might not know how to avoid going over budget. A lot of contractors have trouble keeping their spending within the confines of a budget. When was the last time that you had a job run into higher costs than you were expecting? Do you perform job costs for all of your work? If you're not doing job costing, you can't know how well your estimates worked out with the real cost of a job. It might be surprising to you, but an awful lot of contractors don't maintain regular job-costing procedures.

Before you can stay on budget, you have to create a budget to follow. Far too many contractors are lax in doing a budget. Many contractors work up a lump-sum figure and work only from that number. They might know if they make or lose money, but they don't have enough information to know where the wins and losses come from. This is bad business.

Setting up a job budget

It's not difficult to set up a job budget. You don't need a degree in accounting to manage the task. It does take a little time to establish a framework of prices, but it is time well spent. Most successful contractors never start a job without a budget. You shouldn't either.

When I'm getting ready to bid on a job, I break my bid work down into phases. My breakdowns are done into tight categories. Some contractors lump all of their excavation and grading work into one category. I don't. My procedure calls for a category to meet every major aspect of a job. For example, I would have a category for site clearing. Another category would be reserved for road work. A different category would cover digging footings and foundation holes. Rough grading would have its own section on my budget. So would final grading. Basically, every phase of a job would be broken down. Since we are talking about wells and septic systems, let's use these two aspects of a job as our working example.

Septic systems

The manner you use to break down the costs for a septic system can vary. If you are hiring a subcontractor to provide all labor and materials, you might plug in only one price. This, of course, would be the price quoted to you by your septic installer.

If you are supplying materials for your septic contractor, the cost breakdown would be more extensive. You could lump all material costs into one category, but I'd go further than that. I would list all of the major components of the system separately. For example, I'd have space for crushed stone, concrete products, chambers (if needed), pipe, dirt, and so forth.

My reasons for going into extensive breakdowns are numerous. For one thing, I'm not as likely to overlook an expense if my worksheet details all the major expenses. Using a septic system as an example, I might forget to figure in the cost of a permit if a slot doesn't exist for the price of a permit. It's conceivable that I would fail to figure a price for stone if a blank line or box doesn't exist for the cost. By having a detailed worksheet, I can go through it systematically and not worry about what I might be leaving out. This is good protection against cost overruns.

Once I start a job, I can use my budget to track the profit picture of my work. When an invoice is delivered for crushed stone, I can compare it with the figure I have in my budget. If the stone cost less than what I had figured, I'm in great shape. When the cost is on target, I'm okay. A price that is higher than by bid figure raises a red flag. I can't do much about it on the current job, but it tells me to find out what I did wrong so it won't happen again.

When the dust settles, I can review my estimated figures and compare them with the true costs of materials. It's easy for me to job cost any phase of a job, because I have each phase broken down independently. After a few jobs, I can compare notes and see if I'm experiencing any reoccurring problems. For example, if I'm habitually missing my estimated figures for fill dirt, I know this is an area I need to work on. When I'm constantly coming in under budget on concrete materials, I might decide to lower my prices a little if competition is tough. The budgets that I create can serve many purposes.

Wells

When I have a house coming up that needs a well, I break my budget down into several categories. One of the categories is for the direct well work, either the drilling or the boring. A second category covers the pump. Pressure tanks have a category of their own. Gauges, fittings, pipe, and miscellaneous items are lumped together. Trenching gets its own category. And, of course, labor has a section in the budget. I have used my extensive breakdowns for about 15 years, and I have always found them helpful.

Take-offs

If you are figuring the cost of a job accurately, you must make take-offs. To complete an accurate take-off, you must do extensive breakdowns of the materials and labor that are needed for a particular phase of work. You are probably accustomed to doing this with your building materials. Since wells and septic systems might be beyond your scope of detailed knowledge, preparing a full take-off for such a system might be more than you can handle. If you're using subcontractors for the work, you're not the one that needs to worry about take-offs for wells and septic systems. All you have to do is write or type in the prices given to you by your subs. This makes it all that much easier to come up with budget numbers.

If you are figuring your own materials with a take-off, you can use the take-off as a part of your budget. For example, when doing your take-off, if you decide that 200 feet of PE pipe is needed, you can use that information in breaking down your budget. Putting together a budget from take-offs is simple and effective. You need to create some type of working budget. Take-offs don't have to be a part of this administrative task, but you do need some way to come up with firm numbers. If you cheat and cut corners in order to arrive at your budget figures, you are only hurting yourself. Trying to save time can wind up costing you money—maybe a lot of it.

Accuracy

The accuracy of any budget you create must be dependable. If the numbers aren't solid, they do you very little good. Everyone makes mistakes from time to time. The odds of you making an error in a complex take-off are high, especially if you are not doing the take-off in an undisturbed atmosphere. Even transferring numbers from a sub-contractor's bid package to your budget sheet can create mistakes. Sometimes a seven looks like a one. Forgetting to add a zero to a price can make a huge difference. If you depend on a computerized spreadsheet to add up your figures, you might not notice a simple mistake, until it is far too late. After you have created your budget, go back over it and double check its accuracy.

Sorting through bids

Sorting through bids can be a tedious job. In order to get a clear picture of each bid, it is necessary to spend adequate time poring over proposals and bid packages. During my many years as a builder, I've seen a number of strange proposals and bid packages. Some of the prices I've reviewed turned out to be way out of line. If I had been too busy to study the prices, my negligence could have cost me a bundle. This can be true of prices provided by subcontractors and suppliers.

If you have much experience as a builder, you have undoubtedly had occasions when suppliers have botched up your price quotes. I've had suppliers omit all interior doors from my requests for prices. Hey, if you leave out all the interior doors for a house, your bid price is going to look pretty good. But, a cheap bid price that's wrong doesn't do anyone much good. It's essential that you take an active role in checking and double checking all quotes given to you. Let's put this into perspective for wells and septic systems.

Well prices

Well prices are generally fairly consistent. Some building trades seem to vary in their prices much more than others. Well contractors with whom I've dealt with have normally been close competitors. This makes it easier in some ways, and more difficult in others.

Contractors who install wells typically offer two types of pricing. The options are normally either a per-foot price or a flat-rate fee. Both of these pricing structures have pros and cons. When well installers quote a flat-rate fee, it is usually the highest price an installer expects a well to cost. It is often possible to beat flat-rate fees by choosing a per-foot price. But, picking a per-foot price structure is risky. Flat-rate fees are guaranteed, per-foot prices can run away with your cash.

You must look closely when reviewing bid prices from well installers. Most of the prices are going to be fairly clear and concise. But, there can be hidden charges or wording that might leave you out on a limb. To expand on this, let's start with a per-foot price structure.

Per-foot prices

A quote for a per-foot well does not include a total price. The well installer is going to charge so much per foot for every foot of depth that is required for a suitable well. Casing for the well is also priced on a per-foot basis, with no reference to how many feet are going to be needed. This leaves you with an open-ended deal. You know how much the well is going to cost per foot, but you have no idea of how many feet you are going to be billed for. Under this circumstance, you are at risk if you give a customer a flat-rate quote. Are you going to make your price to the customer based on a per-foot basis, or are you going to make your best guess and give a firm price? Banks and other lenders require a firm price if they are supplying construction financing.

I've only ever gone with flat-rate well pricing once in my long building career. With all the wells I've had installed, I always came out better by gambling on the per-foot price. Then, when I built my most recent personal home, I opted for a guaranteed price. It's a good thing I did. My well turned out to be much deeper than I or my well driller thought it would. To be safe, you should work with guaranteed prices. If you are willing to gamble, you can quote flat-rates to your customers and play the odds of per-foot pricing with your well installer. This can net you more money out of a job, but it can cause you to pay for cost overruns out of your own pocket.

Our main interest right now is in comparing price bids. In the case of well installers, this is pretty easy to do. If all of your well contractors are bidding the same type of well and quoting per-foot prices, you can look at their prices and compare apples to apples. But, there are some opportunities for unethical installers to take advantage of you. They might dig or drill deeper than necessary. The amount of casing installed might not be reflected accurately on your bill. A per-foot pricing basis is similar to a labor-and-material pricing structure. Both are risky. If you like to work with known figures, stick with guaranteed prices.

Guaranteed prices

Guaranteed prices are often available from well installers. These prices tend to be on the high side, but they should be dependable. If you have three well contractors all bidding identical specifications with guaranteed pricing, there is not much effort required in evaluating their prices. All you have to do is look for the low number. In theory, you can't lose money on a guaranteed deal. You can't suffer from cost overruns. If anyone loses, it's the well installer. This approach is a conservative, safe one.

Reading between the lines

Reading between the lines of a quote from a well company is sometimes necessary. Not all quotes are quite what they appear. A busy builder can make assumptions and then find big trouble arising from a lack of facts. Let me expound on this.

Assume that you have received quotes for installing a well. If you're new to wells, you might think that the price you are given includes all the necessary work and materials for a pump system. This is not normally the case. A lot of well installers do install pump systems, but I've never seen one include the pump work in their well price. A well price normally includes nothing more than drilling the well, installing the casing, and capping the casing. I suppose a really slick con artist could try to sell you a well without the casing and cap included. This has never happened to my knowledge, but it won't hurt for you to confirm exactly what is included in all prices you receive.

Pump systems are generally treated as a different job than the well installations. Even when the same contractor is doing both jobs, the work is normally priced separately. It should be. This way you can compare the well installer's pump price with that of your plumbers. If the pump work and well work are mixed together, you can't compare the pump prices supplied by plumbers with those provided by the well installers.

In my experience, grouting has always been included in well installation prices. I suppose this aspect of a job could be left out of a proposal offered by a well installer. The installer might insist that you were planning to grout the well with your own crew. Again, I don't know that this has ever happened, but it is a potential risk.

Your first job as a general contractor is to understand all the bid packages that you get. If you have questions, call the subcontractors and get answers to your concerns. Don't make assumptions, they generally only serve to hurt you. Get all the facts before you award a contract.

All of the well installers I have dealt with in the past have been good ones. I've never had a bad experience with a well installer. I can't say this about many of the building trades I've worked with, but well installers have been good to me. Still, you can't allow your defenses to become too weak. Read each quote carefully. Has anything been disclaimed? It's common for water quantity and quality to be disclaimed, but there shouldn't be any other caveats.

If you are having a well installer bid on your pump package, make sure that all the materials are clearly specified. Don't accept just the name of the pump to be used. Get a model number and description sheet for the pump. Compare flow rates, horsepower, voltage, and other specifics with the pumps that competitive bidders are giving you. Find out the type and capacity of the pressure tank. What type of pipe is going to be installed? Are nylon or brass fittings going to be used? Is the installer going to double clamp all connections? Create a checklist of questions to ask installers. Don't attempt to choose the best bidder until you have all the facts.

Trenching and backfilling are both parts of a pump installation job. Most installers don't include this work in their prices. If you receive a bunch of bids and assume that the trenching is included, you can be in for a rude awakening. The cost of a backhoe or excavator and operator can become quite expensive, especially when it's not in your budget. Confirm who is responsible for digging and filling in the well trench before you make a decision on who gets the job.

Septic prices

Septic prices can involve a lot of material and money. Unlike well prices where per-foot prices are common, most septic systems are priced with flat-rate fees. A septic design is handed out to a number of septic installers and complete prices come back to you. They are generally comprehensive, but there can be some hooks hidden in them.

If the septic system requires a pump system, you need to know who is paying for the pump, the controls, and related pumping materials. Who is going to install this equipment? Are you going to have to pay a plumber to install the pump and an electrician to wire up the system? Not all septic contractors include the price for labor and material that are required to make a pump system operational. Making the mistake of thinking these prices are included in a septic price can cost you thousands of dollars.

Most septic installers provide prices that include all permits, work, and material associated with a designed system. But, you can't count on this. There have been times when I've bought septic permits and septic materials for installers. You have to clarify everything before you commit to accepting a bid.

Excavation can be a big expense when putting a septic system together. This work is normally included in the bids of septic contractors. So is the cost of stone, pipe, and fill dirt. Septic tanks and distribution boxes are commonly provided by septic installers. The prices you get should be turnkey numbers, but check them out.

One aspect of septic systems that does seem to vary is the installation of a sewer between a septic tank and the sanitary plumbing from a house. Sometimes plumbers do this work, and sometimes it's done by the septic installers. This work involves trenching, pipe, fittings, and labor. The amount of money is rarely a fortune, but it's enough that you won't want to pay it out of your own pocket. Find out which one of your subcontractors is bidding the sewer work.

The final grading of a septic system is normally done by the site contractor. This person might also be your septic installer. Many site contractors install septic systems. Again, however, you can't afford to make assumptions. Determine exactly who is going to handle the final grading.

Suppliers

If you are supplying your own materials for either pump systems or septic systems, you are going to need to get prices from material suppliers. This work can be frustrating. Some suppliers don't seem very interested in getting new business.

When I built my most recent home, I sent bid packages to seven major suppliers. Of the seven, only five responded with prices. Two of the suppliers must have decided that my job wasn't worth bidding. Now, I'm not talking about a bid of a few hundred dollars. Tens of thousands of dollars in materials were on my list. Even so, two out of seven suppliers chose not to bid the job. It must be nice to be so independently wealthy that you don't need to bid for work.

Most suppliers bid jobs, but they might not do it the way that you request. I like my bids broken down into itemized lists. A lot of suppliers prefer to throw out a lump-sum figure at contractors. You are at a disadvantage if you don't have itemized pricing. It's possible to look at the bottom-line figures of lump-sum bids and choose a low bidder, but you could be wasting money. Let me explain.

Take a pump installation as an example. If you get three lump-sum bids, you can easily see which supplier is offering the overall lowest price. But, is one supplier more expensive on the well pump than another? Could you buy a pressure tank for less from one place than you could another? You won't know without an itemized list of prices.

It's rare when one supplier offers the best prices on all the materials. This is true of building materials, plumbing materials, and probably most other materials. A supplier who can give you a great price on dimensional lumber might have lousy prices on siding or roof shingles.

Shopping individual prices can save you a lot of money. I'm not one to nickel and dime a supplier, but I do believe in shopping various phases of a job. I might buy a pump from one place, pipe from another, and a pressure tank from yet another. If a couple of extra phone calls can save me a significant amount of money, I'm all for it. Now, I wouldn't buy some fittings here and some there. That gets too confusing. But, within reason, selective shopping of itemized materials is good business.

Some contractors rely on suppliers to do take-offs for them. I don't have a problem with letting suppliers provide a take-off for a job, but I certainly don't trust a take-off that is prepared by a supplier to be accurate. This type of take-off can be quite good, but sometimes it can be way off base. I've seen this problem in both my building and my plumbing business.

Suppliers don't always have the time or the proper personnel to produce quality take-offs. Miscounting a few copper fittings is no big deal, but forgetting to include the cost of a pressure tank is a whole other issue. If you use take-offs prepared by suppliers, I suggest that you check them over very closely. These lists often have omissions.

I generally provide my suppliers with a bid list. After making a take-off, I spec out the brand names I want and ask all suppliers to bid the work as specified. This is the only way I know to get a true comparison of prices. If suppliers are allowed to choose the materials, you might get five quotes and not be able to compare any of them.

When I recommend specifying material, I'm not only talking about big-ticket items. Many contractors spec out a particular pump and pressure tank, but fail to give specifications for valves, fittings, pipe, and other elements of a job. The difference in cost between a cheap gate valve and a name-brand valve can be $15 or more. With enough fittings and accessories, a supplier can lower the cost of a job considerably by pricing materials of a lesser quality. If you don't spec out a job in detail, you won't know what your prices are based on.

Once you get bids from suppliers, read them over thoroughly. I can't count the number of times that suppliers have left whole categories of materials off bids. Go down the bid lists item by item and make sure that everything that you wanted a price on is listed. Simply looking at the total on the bottom of a bid sheet can be very deceiving.

Once a job starts

Once a job starts, you have to stay on your toes to remain within your budget. How often do you compare the monthly invoices from your suppliers with their quoted prices? Ideally, you should check each invoice against quoted prices. Oftentimes, the prices that are charged do not agree with the terms of the quote. How often is this likely to happen? It happens just about every month in my businesses.

When you get a quote from a supplier, it is natural to assume that you are going to be billed at the quoted price. I have found, however, that the prices billed are often higher than the quoted prices. Is this a computer error? Did someone forget to lock in the quoted prices? Do suppliers try to take advantage of contractors? I don't know why it happens, but it does. Sometimes the difference is only a few pennies, but other times the discrepancy is hundreds of dollars.

I try to check each invoice that is tied to a quoted job. Sometimes an invoice slips through without me seeing it, but this is rare. During my checks, I often find problems with my billing. The trouble doesn't come from just one or two suppliers, it seems to happen with a lot of them. There are some suppliers I've never caught in a mistake, but they are the exception, rather than the rule.

Keeping tabs on supplier invoices is a simple way to avoid cost overruns. If a supplier bills you at a price higher than the written, quoted price, you can ask for, and usually receive, the lower price. Even small price increases can add up over the course of a year. Paying $5 too much for this and $10 too much for that, on job-after-job, can cause you to lose thousands of dollars.

It takes time to go over invoices. When the price discrepancies are small, it's hardly worth the time it takes to find and correct them. But, there can be some big differences in prices. On the last house I built, there was a $700 problem with one of my invoices. For this amount of money, I'm willing to scan invoices for errors.

Extras

How do you handle customers who want extras on a job? If someone asks you to add an in-line filter to their job while putting in the pump system, what do you do? A change order should be written and signed by all parties to reflect the change. This is the best way to avoid confusion and possible payment problems. Yet, far too few contractors take the time to use change orders.

It is not uncommon for customers to get additional work done without ever being billed. This is a sure way to bust your budget. If you don't keep good records of job changes, the billing for this additional work can easily slip through the cracks. Mechanics often do little extras on jobs and forget to turn in work orders. Always record any additional work so that it can be properly billed.

Written records

Keeping written records during a job can pay off later. You can use different ways to keep track of what goes into a job. You might just keep delivery tickets and time cards as reference points. Some contractors, like me, use day sheets to list the materials used on any given day. This procedure is more accurate than piling up delivery tickets.

When my crews go into the field for small jobs, they make written records of all materials used on their job. This is done each day that they work on a job. If they work two jobs in the same day, they turn in two different logs. Employee time is also kept in a journal. The time is broken down by phases. If two hours are spent putting in a pump and one hour is spent backfilling a trench, the day-log reflects the work. This type of paperwork makes it easy for me to track production, inventory, estimates, and job costs.

As a job is winding down

As a job is winding down, you should start to gather your job-costing data. This might mean a trip into the field to count the materials used, or you might be able to rely on paperwork created during the course of a job. One way or another, you need to know how much labor and material went into a job. This might be as simple as jotting down numbers from bills you have received from subcontractors. If you are not supplying any of the labor from payroll people and your subcontractors are providing all materials, your paperwork can be kept to a minimum.

By the time a job is finished, you should be geared up to perform a full job-cost report. It might be necessary to wait for the final billing to come in from your suppliers before putting a final job-cost together. This doesn't have to stop you from starting the work. The longer you wait to produce a job-cost report, the more likely you are to neglect doing the report at all.

Pulling it all together

Once you have all the data needed for a job cost, you can start pulling it all together for a clear view of how well you budgeted the job. Only an accurate job-cost report can prove whether you made or lost money. Finding out that you've lost money on a job is never pleasant. But, discovering financial losses after one job can help you to avoid them on future jobs.

Job-costs give you power—profit power. A good job-cost can show you what you did right and where you went wrong. You can determine if your estimating skills are good when you review a job-cost report. There is a wealth of knowledge available to contractors who use job-costing techniques properly.

To stay on budget, you have to track all of your financial activities. This should be done during the course of a job and after a job is finished. Until you can account for all of the costs of doing a job, you can't be sure if your business is profitable. Since profit is a prime reason for being in business, it's very important to keep track of the money you are making or losing.

A money diet

You might have to put your company on a money diet. Sometimes businesses become fat with overhead expenses. This puts strain on a company. Reducing overhead can increase profits. If your business is too fat with overhead, put it on a diet. How does this apply to cost overruns? I'll tell you.

Overhead reduction might not seem to have anything to do with cost overruns on jobs. In fact, it doesn't, assuming that a contractor calculates all overhead into job quotes. Overhead calculations are often done incorrectly, if at all. But, overhead expenses that are not accounted for can ruin the projected profit of a job.

Do you have field supervisors on your payroll? If you do, how do you bill out their time? How many employees do you have? Are some of them administrative employees? Do you account for their cost in your job quotes? How you factor in the cost of employees, insurance, rent, utilities, and other expenses can clearly affect your profit.

Employees cost employers a lot more than their hourly rate of pay. When all associated expenses are factored into a total hourly cost, the cost of an employee can be far more than the hourly rate of pay. Someone who is being paid $10 per hour might be costing an employer $14 an hour. Many factors influence the total cost of employees. Paid vacation, company-provided insurance coverage, and a host of other expenses contribute to the total cost of an employee.

Not all builders have employees. Many contractors work exclusively with subcontractors. When this is the case, figuring overhead expenses is a little easier. I don't plan to get into a long discussion on company overhead. But, I do want you to be aware that any overhead that is not factored into a job quote can cause you to make less money than you planned.

Here we are

Well, here we are, at the end of our time together. You've been given a lot of information on wells and septic systems. I trust you've found it easy to understand and beneficial. Builders, like yourself, who take the time to increase their knowledge tend to be more successful. I wish you the very best of luck in all your endeavors.

Appendix

Resource guide for plumbing products and tools

A. O. Smith
Water Products Co.
P. O. Box 1499
Camden, SC 29020

American Standard Inc.
Plumbing Products Group
One Centennial Plaza
P. O. Box 6820
Piscataway, NJ 08855-6820

Midwest Supply Co.
One South Park Rd.
Joliet, IL 60433

Aurora Pump
800 Airport Rd.
North Aurora, IL 60542-9977

Autotrol
5730 North Glen Park Rd.
Milwaukee, WI

A. W. Cash Valve Mfg. Corp.
666 E. Wabash
Decatur, IL 62525

Central Brass
2950 East 55th St.
Cleveland, OH 44127

Chemical Engineering Corp.
P. O. Box 266
Churubusco, IN 46723

The Chicago Faucet Co.
2100 South Nuclear Dr.
Des Plaines, IL 60018

Conbraco
P. O. Box 247
Matthews, NC 28106

Eljer
3 Gateway Center
Pittsburgh, PA 15222

Fernco
300 S. Dayton St.
Davison, MI 48423

Fiat Products
One Michael Ct.
Plainview, Long Island, NY 11803

General Wire Spring Co.
1101 Thompson Ave.
McKees Rocks, PA 15136

Gerber Plumbing Fixtures
4656 West Touhy Ave.
Chicago, IL 60646

Goulds Pumps, Inc.
P. O. Box 68
East Bayard St.
Senneca Falls, NY 13148

Guy Gray Manufacturing Co.
P. O. Box 2287
Paducah, KY 42002-2287

Kohler Co.
Kohler, WI 53044

Leonard Valve Co.
1360 Elmwood Ave.
Cranston, RI 02910

Little Giant Pump Co.
3810 N. Tulsa
Oklahoma City, OK 73112

Microphor, Inc.
Plumbing Division
P. O. Box 1460
452 East Hill Rd.
Willits, CA 95490-1460

E. L. Mustee & Sons, Inc.
5431 West 164th St.
Cleveland, OH 44142

F. E. Myers
1101 Myers Parkway
Ashland, OH 44805-1923

Price Pfister
13500 Paxton St.
Pacoima, CA 91331

Rain Soft
2080 E. Lunt Ave.
Elk Grove Village, IL 60007

Republic Products
P. O. Box 1010
Ruston, LA 71273-1010

Rheem Water Heaters
5780 Peachtree-Dunwoody Rd. N.E.
Atlanta, GA 30342

The Ridge Tool Co.
400 Clark St.
Elyria, OH 44036

Rockwell International
400 North Lexington Ave.
Pittsburgh, PA 15208

Spartan Tool Division
1506 W. Division St.
Mendota, IL 61342

Speakman
P. O. Box 191
Wilmington, DE 19899-0191

Structural Fibers
920 Davis Rd.
Elgin, IL 60123

Symmons Industries, Inc.
31 Brooks Dr.
Braintree, MA 02184

Thompson Plastics, Inc.
3425 Stanwood Blvd. N. E.
Huntsville, AL 35811

Universal-Rundle Corp.
217 N. Mill St.
New Castle, PA 16103

Vanguard Plastics
831 N. Vanguard St.
McPherson, KS 67460

Water Conditioner, Inc.
509 W. Main St.
P. O. Box 187
Waunakee, WI 53597

Woodford Manufacturing Co.
2121 Waynoka Rd.
Colorado Springs, CO 80915

Zoeller Co.
3280 Old Millers Lane
Louisville, KY 40216

Glossary

ABS Abbreviation for acrylonitrile butadiene styrene, normally used to describe schedule 40 plastic pipe that is black.

accessible Having access to a part of the plumbing system. If you are required to provide access, you can install an access panel or some other device that must be removed before access is gained to the plumbing.

adapter An approved fitting that allows the mating of different types of materials or fittings.

air-break An indirect-waste procedure in the drainage system. The indirect waste enters a receptor through open air.

air chamber A device installed in the potable water system to reduce the effect of air hammer. It gives additional room in the piping system for air to compress and water to expand.

air gap (distance) When used to describe distance, the term *air gap* refers to the unobstructed vertical distance between any device conveying water or waste and the flood level rim at the fixture or other device receiving the water or waste.

air gap (product) When used to identify a product, *air gap* refers to the device used to handle drainage from a dishwasher. It is the item that normally sits near the kitchen faucet and is connected to the drain hose from the dishwasher to the sanitary drainage system.

angle stop Cut-off valve frequently found at plumbing fixtures when the water supply pipes come out of the wall.

anti-siphon Valves or other devices that eliminate the risk of siphonage.

area drain A drain used to receive surface water from grounds or parking areas, for example.

automatic vent A mechanical vent used to vent a portion of the drainage system. These vents are usually made of plastic and contain a diaphragm. The vents are screwed into a female adapter and provide air to the drainage system. As a fixture's drain fills with water, the diaphragm is pulled down to open the vent, allowing air into the drain

273

pipe. This air causes the fixture to drain faster. Automatic vents normally used in remodeling jobs; they are not approved for use without special permission from the code enforcement office.

back vent Vent extending from the drainage of a fixture. A back vent is commonly thought of as a dry vent extending above the fixture's flood level rim.

back siphonage A condition in which water flows backwards in a plumbing system, caused by negative pressure in the pipe.

backflow The reverse flow of water in the plumbing system.

backflow preventer A device used to prevent the danger of backflow.

backwater valve Used in the drainage system, this device prevents sewage from flowing backwards into the building and escaping through the plumbing fixtures.

ball cock Device, operated with a float, used to fill toilets with water.

bell and spigot cast-iron pipe A heavy cast-iron pipe used in the DWV system. This type of pipe has a bell or hub on one end; the other end is straight. To make a connection between two pieces of pipe, the straight end of one pipe is inserted into the hub of the other pipe. The watertight connection is made with oakum and molten lead or rubber devices that fit inside the hub, prior to the insertion of the pipe.

bidet A personal hygiene fixture. While not common in most homes, bidets are found in many upper-end homes.

bleed To purge air from a system. For example, you might bleed the air from the pipes supplying your water pump with water.

branch A pipe that is a part of the plumbing system. Branches originate from the water main or building drain and extend out to a fixture at some distance from the primary pipe. A branch could be any part of the plumbing system that is not a riser, main, or stack.

branch interval A means of measurement for vertical waste or soil stacks. A branch interval is equal to each floor level or story in a building, but they are always at least eight feet in height.

branch vent A vent that connects individual vents with a main vent stack or stack vent. Branch vents can serve a single individual vent or multiple individual vents.

building drain The primary pipe carrying drainage through a building. When the building drain exits the building, it becomes the building sewer.

building sewer Sewer that conveys waste from the termination of the building drain to the connection of the municipal sewer or private waste disposal system.

building trap A trap installed on the building drain to prevent air from circulating between the building drain and sewer.

bushing A fitting that fits inside the hub of another fitting to reduce its size or alter its interior. For example, a 3-x-2-inch bushing would be installed in the hub of a 3-inch fitting to allow the fitting to accept a 2-inch pipe.

cap A fitting placed over the end of the pipe to close the pipe.

caulking (cast iron) The process of making the joints watertight with molten lead.

caulking (fixtures) The act of sealing around the fixtures with a sealant to prevent water penetration.

cesspool A hole in the ground used to accept the discharge of a drainage system. The hole is lined to retain solids and organic material, while allowing liquids to pass through into the earth.

check valve A device used to ensure the contents of a pipe are allowed to flow in only one direction. A common use of check valves includes installation on the pipes of pumps.

circuit vent A vent that serves multiple fixtures. It extends from the low end of the highest fixture connection of a horizontal branch to the vent stack.

cistern A covered container used to store water that is normally nonpotable.

cleanout Accessible opening in the DWV system that allow the cleaning of the pipes with sewer machines and snakes.

closet auger A device used to remove blockages in the traps of toilets. It is a hand-operated tool with a spring head and curved rod, made to negotiate the tight turns of a toilet drain.

closet bend A quarter-bend or 90-degree elbow. Closet bends may have a three-inch opening on one end of the ell and a four-inch opening on the other end of the ell.

code A set of regulations that govern the installation of plumbing.

code officer An individual responsible for enforcing the code.

combination waste and vent system A plumbing system in which few vertical vents are used. In these systems, the drainage pipes are often oversized to allow air circulation in the system.

common vent A vent that serves multiple fixtures.

compression fitting A device using ferrules and nuts on the body of a coupling or fitting to make watertight connections. Compression fittings are frequently found where the supply tubes of faucets enter the cut-off valve.

continuous vent A vent that is a continuation of the drain it serves. Continuous vents are vertical vents and can be referred to as a *back vent*.

continuous waste   A continuous waste is the piping used to connect multiple drains to a common trap. Most double-bowl kitchen sinks are equipped with a continuous waste.

coupling   A coupling is a fitting allowing the connection of two pipes to form a continuous run of piping.

CPVC   Abbreviation for chlorinated polyvinyl chloride plastic pipe, the rigid plastic pipe used in potable water systems.

critical level   The level at which a vacuum breaker may be submerged before backflow occurs.

cross connection   A point at which two separate piping systems, such as the hot and cold water pipes, are allowed to mingle their contents with each other. Cross connections could occur at a faucet, washing machine, or some other location.

crown-vented trap   A trap that has its vent extending upwards from the top of the trap, rather than from the trap arm.

developed length   The measurement or distance of all piping installed. To calculate the developed length of piping, you must measure each piece of pipe.

drain   Any pipe that carries waste in a plumbing system.

drainage fitting   Any fitting used in the drainage portion of a plumbing system.

drainage system   All plumbing that carries sewage, rain water, and other liquid wastes to a disposal site. A drainage system does not include public sewers or sewage treatment and disposal sites.

drum trap   A trap that will not allow the back siphonage of the trap, even without a vent. Drum traps are not self-cleaning and are generally prohibited without special permission from the code enforcement office.

drywell   Sometimes used to receive the discharge from surface-water drains. Drywells are typically lined with stone to accept the water and to allow it to soak into the soil.

DWV   Abbreviation for the drain-waste-and-vent system.

elbow   A plumbing fitting for either water distribution or the DWV system. It may be called an elbow, an ell, a quarter-bend, or a ninety.

escutcheon   A chrome ring found around pipes penetrating finished walls and floors. The escutcheon provides a neat and finished look to the installation of pipes, and they prevent rodents from climbing the pipes to enter the premises.

faucet   Device that controls the flow and mix of water entering a sink, lavatory, tub, shower, or other plumbing fixture.

female connection   A connection with interior threads designed to accept the external threads of a male adapter.

filter Device used to filter substances. The aerator on your faucet is a form of filter.

finish plumbing The setting of fixtures, like toilets, sinks, and faucets.

fittings The parts of the plumbing system used to connect the piping.

fixture branch Water supply pipes running between the primary water pipes and the supply tubes of individual plumbing fixtures.

fixture drain Section of piping that runs from the fixture's trap to the connection with the DWV system. Fixture drains are often called trap arms.

fixture supply Pipe that runs between a fixture and the fixture branch.

fixture unit A unit of measure used to calculate the load and demands of fixtures on a plumbing system.

flange A mounting surface, for example, closet flanges, which are the devices mounted to the floor to receive and hold toilet bowls.

flapper Rubber device used to seal the opening of a flush valve in a toilet tank.

flare fitting A fitting made to be used with flared piping. Flare fittings are specially formed to mate with flared pipe in creating a leakproof joint.

flashing Devices used to seal around vent terminals on the roof of the structure.

flex connector A short piece of material used to connect a device with the piping of a plumbing or gas piping system. Flex connectors are flexible and allow the device connected to the rigid piping to move without putting stress on the rigid piping.

float ball The ball at the end of the float rod in the toilet tank, which operates the ball cock.

float rod A rod between a ball cock and a float ball. The float rod screws into the ball cock and the float ball to allow proper operation of the ball cock.

flood-level rim The point at which water will flood out of a fixture. For example, the edge of your kitchen sink, where it meets the countertop, is the flood-level rim of the sink.

flush valve Device in a toilet tank that allows water to pass from the toilet tank to the toilet bowl.

flux A substance applied to copper pipe when the pipe will be soldered. The flux acts as a cleaning agent to ensure a good solder joint.

gate valve A valve that uses a forged, metal gate to close the valve. Instead of using a rubber washer, which may deteriorate, a gate valve closes with the use of the gate, ensuring a more positive closing of the valve.

grade   The slope, fall, or pitch of a pipe.

groundworks   Plumbing installed below a finished grade or floor.

hose bibb   Device used to supply water to garden hoses. A hose bibb accepts a water supply pipe on one end and the threads of a garden hose on the other end. They are equipped with a handle and assembly to control the flow of water.

hot water   Water is considered to be hot water when its temperature is at or above 110 degrees F.

house trap   House traps are not allowed by present plumbing codes, but they can still be found in older existing buildings. They are a trap installed in the building drain, just before the building drain exits the building.

hub   The part of a pipe or fitting designed to accept the end of a piece of pipe.

hubless cast iron   Pipe that is a lightweight cast iron without hubs. The connections made with this type of pipe are made with special bands and clamps.

hydrostatic test   A test of the plumbing system using water. If you test your waste or water lines with water, you are performing a hydrostatic test.

indirect waste pipe   A pipe that does not connect directly to the sanitary drainage system. Instead, it conveys its contents to the sanitary drainage system through an air gap to a fixture or other receptacle.

individual vent   A vent serving only one trap.

J-bend   The portion of a trap that retains water at all times.

joint   A connection made in the installation of plumbing.

long-sweep fitting   A fitting with a more gradual turn than its short-turn counterpart. The more gradual bend reduces the risks of pipe blockages.

main   A primary pipe, such as a water main.

main sewer   The public sewer.

main vent   The primary vent for a plumbing system.

male adapter   A fitting with exposed threads, designed to screw into a female adapter.

nipple   A short piece of pipe with threads on each end.

nonpotable water   Water that is not safe to drink.

O-ring   A rubber ring used in plumbing parts to seal them against water leakage.

oakum   A material packed in the hub of cast iron pipe that surrounds the pipe extending into the hub, before molten lead is poured into the hub to make a joint.

offset A change in direction. Any time your piping is turned in a different direction with the use of fittings, you are creating an offset.

P-trap The most common type of trap used in modern plumbing. It is used for trap arms that come out of the wall, as opposed to the floor.

packing nut The nut in a valve that holds the packing in the valve to prevent leaks.

PB pipe Polybutylene pipe, a very flexible plastic pipe used for water distribution.

PE pipe Polyethylene pipe, a plastic pipe that comes in rolls and is frequently used for water service applications outside the building.

pipe joint compound Also called pipe dope; a sealant applied to threads before making a screw connection.

pitch See *grade*.

plug Similar to a cap, except a plug is screwed into fittings instead of installed over pipes.

plumbing The trade or work pertaining to the installation, repair, alteration, and removal of plumbing and drainage systems.

plumbing code A set of rules, regulations, and or laws that dictate the manner in which the trade of plumbing may be conducted.

plumbing official Also called a plumbing inspector; the authorized individual designated to inspect and enforce the plumbing code.

plumbing system All water distribution pipes, waste and vent pipes, plumbing fixtures, traps, and devices used for plumbing within the property lines of the premises.

potable water Water that is safe for drinking, cooking, and domestic use.

pressure pipe A pipe meant to handle contents that are under pressure, such as a water distribution pipe.

pressure-reducing valve A valve that governs the pressure of a substance entering a pipe. The most common use of pressure-reducing valves is their use on the water main of a building. If the water pressure produced by a municipal source is too high for the building's use, a pressure-reducing valve is installed to lower the pressure entering the building's water distribution system.

private sewage disposal system A sewage disposal system serving a private party. A septic system is an example of a private sewage disposal system.

private water supply A water supply serving a private party, such as a well.

PVC An abbreviation for polyvinyl chloride plastic pipe, frequently used in DWV systems. When used for this purpose, PVC pipe is schedule 40 plastic pipe that is white.

readily accessible Having direct and immediate access to an object. If an access panel must be removed before the object can be accessed, the object is not readily accessible.

receptor An approved device intended to accept the discharge from an indirect waste.

reducer A fitting that reduces the size of a pipe or fitting. Unlike bushings, reducers are installed over pipe, rather than in the hub of fittings.

relief-valve drain The drain that handles the discharge of a relief valve. These drains usually terminate into open air, not into the sanitary drainage system.

rim The open edge of a fixture.

riser A pipe that rises vertically.

rough-in The plumbing installed prior to the completion of walls, ceilings, and floors. If you were plumbing a new house, the plumbing you installed before drywall was placed on the walls and ceilings would be rough-in plumbing.

S-trap Now illegal, these traps were used on fixtures when the drain for the fixture comes up through the floor, instead of coming out of a wall.

saddle valve A valve used to tap into existing pipes. A common use of a saddle valve would be to supply an ice-maker with water from the supply pipe of a kitchen faucet.

sanitary fitting A fitting used in drainage piping.

sanitary plumbing The plumbing that removes waste from a building.

septic tank A part of a private waste disposal system, septic tanks accept the discharge from a building sewer and hold it for distribution into the septic field.

sewage Any liquid waste containing animal or vegetable matter.

slip fitting A fitting that slides over a pipe and continues to slide down the pipe. Slip couplings are the type of slip fitting most often used. They allow the coupling to slide back onto the pipe so that the couplings can be used with a minimum of space. By using slip couplings, you do not need to gain much movement of the pipe to make a joint.

slope See *grade*.

soil pipe A pipe that transports sewage containing fecal matter.

soil stack A soil pipe that extends vertically to accept the discharge of toilets.

solder A substance used to make watertight joints with copper pipe and fittings.

soldered joint A joint made with solder.

solvent cement The glue used to make connections with various plastic pipes.

stack A vertical pipe in the drainage system. A stack may be a vent, soil pipe, or waste pipe.

stack vent A vertical pipe that extends above the highest drainage point to vent the drainage system.

standpipe A receptor for the drainage conveyed by washing machines. They are the vertical receptor extending from a trap to accept indirect waste.

street fitting A fitting that is made so that one end of the fitting will fit into the hub of another fitting.

sump A tank or basin that receives sewage or water to be pumped to another location.

sump pump Used to pump water from sumps; not intended to pump sewage.

sump vent When a sump is used to receive sewage, the sump should have a sump vent. The sump vent controls odor and sewer gas that builds up in the sump.

sweating Condensation.

sweating (soldering) The act of soldering. Plumbers often say they are sweating the copper pipes and fittings to mean they are soldering the pipe and fittings.

T & P valve Temperature and pressure-relief valves. These valves can be found on water heaters. They are a safety device to protect against excessive heat or pressure in the water heater.

tailpiece The tubing that extends from a fixture's drain to the fixture's trap.

tank ball Used to seal the openings of some flush valves in toilet tanks.

tee A fitting that allows a secondary pipe to branch off of a main pipe.

tempered water Water tempered to maintain a temperature between 85°F and 110°F.

test tee A special fitting used in the DWV system. These tees have a flat face with the tee portion being threaded to accept a clean-out plug. Test tees allow the testing of the DWV system when it has already been connected to the main sewer.

trap arm The section of drainage pipe that runs from the trap to the connection with the building's DWV system.

trap seal A seal made by water that remains in the trap at all times. Trap seals prevent sewer gas from entering the atmosphere inside the building.

trap A device used to prevent sewer gas from entering the atmosphere of a building.

tubing Small, usually flexible, piping. Tubing can be rigid. Technically, most copper pipe used to plumb a home is tubing. It is rigid tubing, but most people refer to it as pipe.

underground plumbing See *groundworks*.

union A fitting used to couple two pipes together. Unions are machined to fit together without leaking. When unions are installed, the connection can be loosened and the two pipes can be separated without cutting the pipe.

vacuum A pressure less than that exerted by the atmosphere.

vacuum breaker A fitting that protects against backflow on openings under normal atmospheric pressure.

vent pipe A pipe used to vent a plumbing fixture.

vent stack A vertical vent pipe connected to the drainage system.

vent system A system of vents designed to provide air circulation to a drainage system and to prevent the siphonage of traps.

waste pipe A pipe conveying waste without fecal matter.

waste The discharge from a plumbing device that does not contain fecal matter.

water closet A toilet.

water distribution pipe A pipe that distributes potable water to plumbing fixtures.

water hammer A pressure surge within the water distribution system.

water-hammer arrestor A device that defeats water hammer by absorbing pressure surges.

water heater A device used to heat cold water to a temperature of at least 110 degrees F.

water main A water supply pipe for public use.

water meter A device installed on a water service to measure the amount of water used by a plumbing system.

water service The pipe delivering water from a water source to the water-distribution system of a building.

water softener A device used to condition water and to remove hardness from water.

water supply pipe Also called a water service, it is the pipe supplying water from the main source to a building.

water supply system All the pipes and parts necessary to supply water to a building.

wax ring   A ring used to create a seal between a toilet bowl and its flange.

well   A source of water.

wet vent   A pipe that receives the drainage from plumbing and does double duty by venting part of the plumbing system.

wing elbow   Also known as a drop-eared ell, it is an elbow with two ears on it. The ear extensions allow the ell to be secured with screws or nails to wood blocking. The most common use of these fittings is to accept the end of a shower head's arm.

wye   A fitting that allows a branch pipe to enter a primary pipe at a gradual angle.

Index

About the author

R. Dodge Woodson has nearly 20 years of experience as a home-builder, contractor, master plumber, and real estate broker. He is also the author of many books from McGraw-Hill, including *Builder's Guide to Residential Plumbing, Home Plumbing Illustrated, Plumbing Contractor: Start and Run a Money-Making Business, Troubleshooting & Repairing Heat Pumps, National Plumbing Codes Handbook, Professional Modeler's Manual: Save Time, Avoid Mistakes, Increase Profits*, and *Master Plumber's Licensing Exam Guide*.